万达商业规划

持有类物业　　下册 VOL.2

WANDA COMMERCIAL PLANNING 2014
PROPERTIES FOR HOLDING

2014

万达商业规划研究院　主编

中国建筑工业出版社

CONTENTS
目录

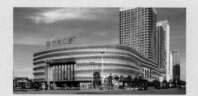

FOREWORD
序言

WANDA
COMMERCIAL
PLANNING
2014

CHAIRMAN WANG JIANLIN TALKING ABOUT WANDA'S EXECUTION

王健林董事长谈万达执行力

万达在全球首创了慧云智能系统。过去购物中心的管理与监控都是分成若干个系统，机电管机电，消防管消防，节能管节能，每个专业都单独监控，相互之间不联通，这种方式既浪费人力资源，而且无法完全避免人所犯错误。这方面万达曾有过深刻教训，这也促使我们想尽办法避免事故出现。经过多年研发，万达2013年成功开发出一套系统，我给它起名叫慧云，寓意智慧的云，就是把万达广场、万达酒店中的消防、水暖、空调、节能、安全等所有监控系统集中在一个超大屏幕上，完全智能化监控。比如员工值班，临到换班时，他会自动向交接班人的手机发短信，提醒按时换岗。比如空调，系统如果检测到某个区域人少，会自动减少送风量，起到节能效果。目前这套系统已在四个万达广场进行试点，今年准备在全国万达广场、万达酒店全面推广。

万达靠制度、文化、科技等综合因素形成了不敢说世界第一、但至少在中国第一的企业执行力，执行力是万达取得今天成绩的秘诀之一。

——摘自王健林董事长2014年4月12日在中欧商学院所作《解密万达执行力》主题演讲

万达集团董事长
王健林

Huiyun Intelligent System is a management system firstly developed in the world by Wanda Group. In the past, the management and monitoring of shopping malls are divided into several systems, in which specialties like M&E, fire protection and energy saving are in the charge of different departments under separate monitoring and communications between the different specialties are not available. Such method not only wastes human resources, but also leaves room for human errors. In this respect, Wanda has learnt its lessons, which urges us to come up with solutions to avoid accidents. After years of research and development, in 2013, Wanda successfully developed a system, which I named "Huiyun", meaning "Intelligent Cloud". The system is able to integrate all the monitoring systems for fire protection, water heating, air conditioning, energy saving, security, etc. of Wanda Plazas and Wanda Hotels into one over-large screen display, making complete intelligent monitoring a reality. For example, before the time for the next shift, the system will automatically send a message to the telephone of the next person on duty to remind him to go to work on time; take air conditioning for another example, if the system detects that the number of people in a certain area is relatively small, it will automatically reduce the air output to such area to save energy. Currently, the system is under pilot operation in four Wanda Plazas, and we prepare to implement it to all the Wanda Plazas and Wanda Hotels nationwide this year.

With an integration of well-established system, inspiring corporate culture and advanced technology, Wanda has achieved an excellent execution, which if not ranks first in the world, at least ranks first in China. Execution is definitely one of the secret for Wanda's achievement today.

——Excerpted from Chairman Wang Jianlin's keynote speech titled Decode Wanda's Execution delivered in China Europe International Business School on 12th April, 2014

Chairman of Wanda Group
Wang Jianlin

PLANNING DESIGN AND CONTROL UPGRADING
规划设计与管控提升

总裁助理兼万达商业规划研究院院长　叶宇峰

万达规划作为万达集团的核心部门，对万达集团的持有物业进行全专业、全过程技术管控。随着集团第四次转型的深入开展，万达规划如何在集团飞速发展的进程中跟上时代的步伐，符合集团转型要求，成为万达集团新的核心竞争力，将是规划院今后工作的重点！

为此，我们要转变思想，不要墨守成规，在集团互联网创新总体要求下，从管理上创新，开拓思维，为集团贡献更多创新设计产品而努力。为此，规划院结合自身业务特点，为进一步提高工作效率，提升业务创新水平，将重点在"设计标准化"、"设计模块化"、"设计创新化"、"设计总包化"和"设计产业化"五个方面做好工作。

一、设计标准化

将"轻资产"项目分为B、C、D版，结合地上、地下不同建筑面积，三层、四层街等不同组合，研发出具有典型意义的标准化产品；同时，加快推进外立面、室内装饰、夜景照明、景观环境、机电设备、基础结构等标准化设计（图1）。

二、设计模块化

运用设计标准化成果，通过互联网思维，将设计产品进一步模块化分类；同时运用计算机软件技术大大提高前期概念方案和单体方案的设计效率（图2）。

三、设计创新化

运用大数据、慧云系统，直接面对市场、数据，真实地反映终端消费者的需求。根据消费者对空间、动线、业态、审美、舒适度等的需求反馈，设计出符合集团利益最大化、符合城市定位、符合项目定位、符合消费者需求的创新设计产品。

四、设计总包化

为配合"施工总包，交钥匙"管理模式，规划院将重点培养若干支服务意识、技术能力过硬的优秀设计供方

As one of the core departments of Wanda Group, Wanda Planning Institute is responsible for the technical control of all the properties owned by Wanda Group concerning all specialties in the whole process. Following the deepened implementation of the Group's fourth transition, it has become the key points of the Planning Institute's future work to keep pace with the times in the progress of the Group's rapid development, to satisfy the Group's transition requirements and to become the new core competitiveness of Wanda Group.

For this purpose, we shall start to change our thoughts and not stick to the accustomed rules. Under the general requirements of Internet innovation proposed by the Group, we shall work hard to produce more creative design products for the Group by means of management innovation and creative thinking. To this end, combining our own business features, to further improve work efficiency and promote the ability for business innovation, the Planning Institute lays emphasis on the excellent execution of five tasks, which respectively are design standardization, design modularization, design innovation, design industrialization and the implementation of design main contract model.

I. DESIGN STANDARDIZATION

The standards for asset-light projects shall be divided into Edition B, Edition C and Edition D. In combination with different areas of aboveground construction and underground construction and various combination of three-storey and four-storey types, typical and standardized products shall be developed; in the meanwhile, the standardized design of façade, interior decoration, nightscape lighting, landscape, electromechanical device, foundation structure, etc. shall be accelerated and propelled (see Fig.1).

II. DESIGN MODULARIZATION

With the utilization of the achievements of design standardization, under the guidance of Internet thinking, the design products shall be further modularized; meanwhile, with the application of computer software technology, the design efficiency of concept plan and individual plan in the initial stage shall be greatly improved (see Fig.2).

III. DESIGN INNOVATION

With the utilization of big data and the Huiyun System, the demands of end customers shall be reflected truthfully by facing the market and statistics directly. Based on the feedbacks from customers concerning the requirements on space, circulation, types of structures, aesthetics, comfort, etc, creative design products that maximize the Group's profit and accord with city orientation, project orientation and customer demands shall be designed.

（图1）《轻资产万达广场机电专业屋面设计标准》

（图2）由万达商业规划研究院开发的"慧图"系统界面

（图3）万达规划研究院举办产业化论坛

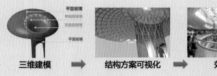

三维建模 → 结构方案可视化 → 效果图

（图4）BIM设计应用实例：汉街万达广场

IV. IMPLEMENTATION OF DESIGN MAIN CONTRACT MODEL

In coordination with the construction "Overall Contract, Turnkey Contract" management model, the Planning Institute will pay attention to the cultivation of several great design providers with strong service awareness and excellent technical competence; the design of a large number of asset-light, asset-heavy, cultural tourism projects and overseas projects will be outsourced under Main Contract model; third party will be brought in to check drawings; more efforts will be put into the cultivation of a group of outstanding main contract and sub-contract design providers; objects of enhancing work efficiency, reducing control costs and upgrading design quality shall be achieved.

V. DESIGN INDUSTRIALIZATION

Through design industrialization, the high-speed duplication, rapid assembly and high-quality completion of Wanda's asset-light projects on a large scale shall be achieved. How to deal with the increasingly huge amount of asset-light projects under the "Overall Contract" model? The Planning Institute initiates the concept of Wanda Plaza's "industrialization" (see Fig.3).

In accordance with the principle of from easy to hard and taking the experience gained at one unit and popularizing it into other units, with the cooperation offered by the design, main contract and research departments, the Commercial Planning Institute, taking "day-lighting roof" as a starting point, devotes itself to the study of standardized design of day-lighting roof and the exploration of an operation mode for the centralized purchasing, factory manufacture and onsite assembly of day-lighting roof. The purposes of which are to achieve a large-scale, low-cost and effective application of day-lighting roof, and to yield the economic benefits brought forth by large-scale production and intensification management of day-lighting roof. The ultimate purposes of which are to cut down design costs, to reduce onsite workload, wet construction and pollution and to ensure the completion of construction on time with high quality and efficiency for Wanda Plaza contents that can be industrialized, and to help achieve the Group's objective of opening 500 Wanda Plazas by 2020.

On the premise of "Internet plus" thinking, in combination with BIM whole-process information management, the Planning Institute has carried out initial exploration in aspects of cloud data planning and design, and research of design intellectualization to further improve work efficiency and save time, manpower and costs (See Fig.4).

In 2015, Wanda will further implement "Overall Contract, Turnkey Contract" model and lay more emphasis on asset-light products, in addition, the Group will try to transform from a pure properties enterprise to a high-tech enterprise of service industry, all of which indicate that the Group is like a high speed train spurring on the railway of "transition". Along with the internationalization and diversification process of the Group, under the strong and brilliant leadership of the Group's leaders, the Commercial Planning Institute will keep up with the time and the development step of the Group. With vigorous passion, positive spirit, sharp mind and innovation awareness, arduous efforts will be made to forge the Planning Institute into a top-level, tough and down-to-earth research and development center characterized by Internet planning and innovative thinking.

队伍；将大量"轻资产"、"重资产"和文旅项目、境外项目进行设计"总包"；引进第三方审图；加大力度培养出一批优秀的"总包"、"分包"设计供方单位；达到提高工作效率、减少管控成本、提升设计质量的目标。

五、设计产业化

通过设计产业化，使万达大规模"轻资产"项目高速复制、快速装配、保质完成。如何在"总包"模式下应对规模日益庞大的"轻资产"项目？商业规划院首次提出了万达广场"产业化"概念（图3）。

根据从易到难、由点到面的原则，商业规划院在设计、总包和科研等单位的配合下，以"采光顶"为突破口，着力研究采光顶的标准化设计，探索采光顶的集中采购、工厂加工、现场拼装的操作模式。目的是实现采光顶的大规模、低成本、高效率的应用，发挥采光顶规模化、集约化的经济效益。最终目的，是使万达广场可产业化的内容实现降低设计成本、减少现场工作量、减少湿作业、减少污染、保质高效按时建造完成，配合集团2020年实现500个万达广场的目标。

在"互联网+"思维的前提下，结合BIM全程信息化管理，规划院在规划设计云数据、设计智能化研究等方面进行初步探索，以进一步提高工作效率，节省时间、节省人力、节省成本（图4）。

2015年，从"总包交钥匙"到"轻资产"模式，从"去房地产化"到高新科技服务业企业，集团正如一部高速行驶的列车在"转型"大道上飞奔！伴随着集团的国际化、多元化进程，商业规划院将在集团英明领导层的坚强领导下，跟上时代的潮流和企业发展的步伐，以饱满的热情、高昂的斗志、敏锐的头脑、创新的意识，为实现将规划院铸造成一支能打硬仗、接地气、具有互联网规划创新思维的行业顶级研发中心而努力奋斗！

WANDA
PLAZAS
万达广场

WANDA
COMMERCIAL
PLANNING
2014

北京通州万达广场
南宁青秀万达广场
兰州城关万达广场
龙岩万达广场
广州番禺万达广场
烟台芝罘万达广场
江门万达广场
福清万达广场
温州平阳万达广场
杭州拱墅万达广场
银川西夏万达广场
满洲里万达广场

BEIJING TONGZHOU WANDA PLAZA
北京通州万达广场

时间 2014 / 11 / 29　**地点** 北京
占地面积 10.39 公顷　**建筑面积** 58.99 万平方米

OPENED ON 29th NOVEMBER, 2014　**LOCATION** BEIJING
LAND AREA 10.39 HECTARES　**FLOOR AREA** 589,900 M²

OVERVIEW OF PLAZA
广场概述

北京通州万达广场位于通州区北苑商务区内黄金地段，东至通惠南北路，南至现状住宅小区，西至北苑南路，北至新华大街。广场占地10.39公顷，总建筑面积58.99万平方米。其中A、B区为商业片区，用地6.49公顷，A、B区建筑面积46.68万平方米。

Located at the prime location of Beiyuan Business District in Tongzhou District, Beijing Tongzhou Wanda Plaza faces Tonghui North-South Road to the east, existing residence community to the south, Beiyuan South Road to the west and Xinhua street to the North. It has a land area of 10.39 hectares and total floor area of 589,900 square meters, including the commerical areas A and B built on the 6.49 hectares land, with the floor area of 466,800 square meters.

02

FACADE OF PLAZA
广场外装

历史中，通州是围绕大运河上航运贸
易和渔业资源生息繁衍的水陆码头。
通州万达广场立面设计的灵感就来自
于"流动"。基于风、水、交通等肌理
流动的概念，将建筑巨大的体量雕刻
成一个曲线船体的形状。这个建筑塑
型与流体力学中"速度"对船体形状的
塑造极其相似，而速度这个概念也与
通州的高速发展相暗合。

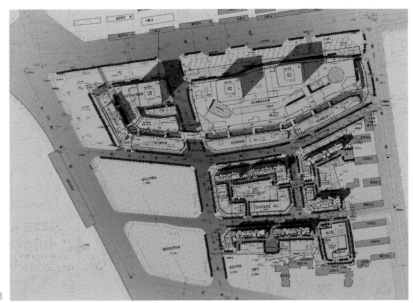

03

04

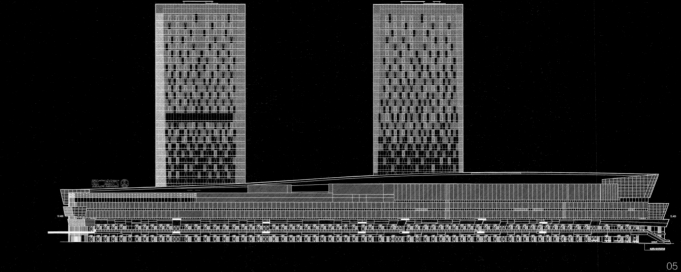

05

06

07

In ancient times, as a land and water wharf, Tongzhou was developing and thriving by virtue of shipping trade and fishery resources. Facade design idea for Tongzhou Wanda Plaza is conceived from "flow". In light of texture flow concept of wind, water and transport, the large volume building is designed as a curved ship, extremely similar to the ship shape formed by "speed" in hydromechanics. Meanwhile, the concept of "speed" coincides with rapid development of Tongzhou district.

08

09

10

11

08 广场外立面特写
09 广场外立面特写
10 广场外立面
11 广场百货入口

商业裙房表面波动的建筑表皮产生出动态的、有活力的城市景观，引起行人的关注，将客人从建筑周边吸引到主入口来。表皮不规则的形态刻意定义出公共和私有空间的层次。当表皮褶皱翻转到两个中庭之间时，金属壳的平滑流动就将室内空间和室外环境的差别抹去了，保证了流动概念的无缝延续。

The fluctuant building surface of the commercial podium has produced a dynamic and energetic landscape, attracting the surrounding crowds' to the main entrance of the building. The irregularly-shaped building surface purposefully defines the levels of public and private space. When the surface folds overturns between the two atriums, the difference between the interior space and exterior surroundings can be erased by smooth flow of the metal shell, resulting in seamless continuation of the flow concept.

INTERIOR OF PLAZA
广场内装

中庭部分的白色环形设计，使人们对店铺一侧和公共空间一侧的感觉具有明确的区分效果；从中庭处向上看时，整个空间生产宏大的连续感及富于冲击力的造型感。靠近店铺一侧的部分天花面上使用黑色的基调，与店铺里的"花哨"产生对比。为了强调主动线，在入口处、滚梯位置以及电梯位置的中庭，特意增加了具有方向指示性的格栅造型。格栅造型的设计，本身具有装饰性同时具有功能和指向性。

By adopting white circular design in atrium, people can clearly distinguish between the store side and the public space side. Looking up at the atrium, the whole space presents the sense of grand continuity and impactful modeling. Partial ceiling close to store largely in black creates a striking contrast with store's showy decoration. In order to highlight the main flow line, the directional grille modeling is expressly added to the atriums at entrance, escalator and elevator. Virtually, the grille design itself has both decorative and functional directivity functions.

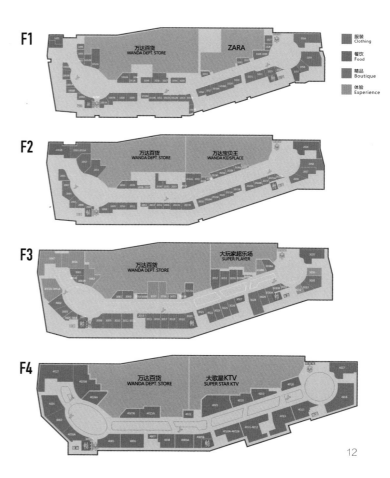

12

13

12 商铺落位图
13 椭圆中庭穹顶
14 椭圆中庭全景

14

15

通过公共空间的简洁设计，突出不断变化以及具有多彩性的店铺，以此对比搭建起自己的地位和风格。公共空间的设计，达到了对各种色彩以及购物中心各种信息的整合使整体空间更具有秩序性的效果。

自2013年起，万达集团开启了扶持大学生创业的十年计划。2014年11月，在北京通州万达广场开业的"鲸品披萨店"便是在这个计划的扶持下由北京大学学生创业团队开创并持有的一家品牌。

By adopting the idea of concise design and highlighting the constantly changing and colorful stores, the public space builds its own status and style through the comparison. The public space has integrated all kinds of color and various information of shopping center, making the whole space more orderly.

Since 2013, Wanda Group has launched the ten-year plan to support university student's innovative undertaking. In November 2014, as a brand founded and held by student's entrepreneurial team of Peking University, "King's Pizza", benefiting from the plan, opened at Beijing Tongzhou Wanda Plaza.

15 公共空间分析图
16 椭圆中庭侧裙板
17 大学生创业店铺店面
18 大学生创业店铺内景
19 美陈特写
20 导识特写
21 室内步行街

17

18

16

19

20

直街悬挂的艺品"形容"了运河的流动,并给空间带来动感,可从不同角落发现不同的美感。天花采用格栅"描述"了整个建筑的流线,并在黑色格栅的背景中商铺门面自然显现出来。格栅造型的设计,在兼具装饰性的同时具有功能性的指向性,提升了商业空间的识别性。

连桥的不锈钢水纹在墙壁及地面投射出如同水面纹路的光影,仿佛水脉流向四面八方。为了更好地传达商业信息,在必要的位置设置了LED照明、电子板等先进的设备。

The straight gallery hangs various artworks to present the flowing canal and add the dynamic sense of space, so that the varied aesthetics can be felt at various corners. The ceiling adopts the grille to present streamlines of the overall building and stores are naturally appearing in the black grille background. The grille modeling is designed with the character of decoration as well as functional directivity, promoting identity of commercial space.

Stainless steel water ripple of connecting bridge projects light & shadow on the wall and ground, looking like water surface texture, as if the water flows in all directions. In order to better convey business information, advanced equipments are placed in necessary location, such as LED lighting and electron plate.

21

WANDA DEPT. STORE
万达百货

北京通州万达百货坐落于占地面积10.39公顷的通州万达广场内，营业面积1.3万平方米，共5层；定位于时尚精致生活，辐射整个通州区域。

通州万达百货设228个柜位，其中A级品牌占有率达到46%；在重视国际品牌对百货档次提升的同时，也不忽视居民日常生活中不可取代的传统品牌和民族品牌，成为通州区最大的综合型购物商场。各楼层业

态布局合理，标杆品牌市场影响力较大；适卖品牌比例较高，符合区域顾客的需求。

连通的品牌装修与步行街的品牌装修无缝衔接，边厅品牌形象装修到顶，专柜与专柜之间的衔接自然、合理；全店色彩使用鲜明，各楼层风格结合软装、美陈与绿植，将高端专业的卖场氛围与自然的清新舒适融为一体，打造出一流的购物天地。

23

22 万达百货街景
23 万达百货 "一店一色"
24 万达百货 "一店一色"

24

The five-storey Beijing Tongzhou Wanda Department Store is located at Tongzhou Wanda Plaza, occupying the land area of 10.39 hectares and the business area of 130,000 square meters, aiming at fashionable and exquisite life and covering the whole Tongzhou district.

Tongzhou Wanda Department Store accommodates 228 counters, of which Class A brand share reaching 46%. While upgrading the commodity grade through international brands, irreplaceable traditional and national brand affections are highlighted as well, which make it become the largest comprehensive shopping center in Tongzhou. With reasonable business type distribution on each floor, benchmarking brand occupies an influential market. With higher proportion of saleable brands, the local customer demands are satisfied.

Seamless connection is achieved between the connected branda decoration and brand decoration in pedestrian street; brand decoration in side hall reaches the top; and natural and reasonable connection is available between counters. By using bright color in store, integrating soft decoration, display and green plants into each floor, and fusing high-end and professional store atmosphere and naturally fresh and comfortable perception together, a world-class shopping center is presented.

WANDA CINEMA
万达影城

通州万达影城位于万达广场的五层，是一个全开放空间的巨大交流空间，为万达院线中的高端级别，文化气息浓郁，"展示影院2.0时代"的突出特征：（1）可以体验多元消费；（2）首推万达院线的自有餐饮品牌"ONE"；（3）遍布WIFI SPACE；（4）体现"电影院应是最体现电影文化的商业空间"；（5）体现"科技，环保，人文"的理念；（6）观影的功能得到拓展，可以举办多种活动的功能；（7）强化了文化的功能，可以举办艺术巡回展；（8）设置一座400平方米的屋顶露台，具有多种商业功能。

Located on the fourth floor of Beijing Tongzhou Wanda Plaza, Tongzhou Wanda Cinema is a high-end product in Wanda Cinema Line with a gigantic and fully open communication space. Tongzhou Wanda Cinema is prominently characterized by strong cultural ambience and "Exhibition of Wanda Cinema 2.0": first, various consuming behaviors are allowed; second, the proprietary catering brand of Wanda Cinema Line- "ONE" enjoys its debut here; third, WIFI SPACE spreads all over the cinema; fourth, it reveals the design philosophy of "Cinema shall be a commercial space that best demonstrates movie culture"; fifth, it reflects the romanticism of "Technology, environment protection and culture"; sixth, the functions of the cinema are extended, allowing for the holding of multiple events; seventh, cultural function is extended, allowing for the holding of art exhibition tour; eighth, a 400 square meters roof terrace with multiple commercial functions is set.

26

THE SUPERSTAR
大歌星KTV

通州万达大歌星位于广场1号门4楼，总面积3395平方米，包厢数55个，其中包括总统、商务、VIP、主题包、中包、小包等。全场无线麦克风，复古立麦，不同的装修风格能满足顾客的不同需求。店内风格以紫色系列为主，打造梦幻空间感。各类包房装饰风格丰富多彩、豪华典雅，全方位满足顾客欢唱、畅饮需求。配置顶级高端的硬件设备，完美音质的立体麦克风、热力四射的表演舞台。大歌星及时更新歌库，热门歌曲全部覆盖。

Tongzhou Wanda KTV is located at 4F No.1 Door of Wanda Plaza, covering the total area of 3,395 square meters and 55 boxes, including president, business, VIP, theme, medium and small boxes. It is equipped with wireless microphone, retro stand microphone and decorated in various styles, catering for the demands of various customers. Adopting purple series in style, the KTV creates a fantastic space. With rich, colorful, luxurious and elegant decoration styles, the various boxes enable to meet the demands of singing and drinking. The Singer is equipped with advanced hardware equipment, stereo microphone with perfect sound quality and dynamic performance stage. Moreover, the song-list is timely updated to collect all hit songs.

30 大歌星特色包房
31 大歌星特色包房
32 大歌星总统包房
33 次主力店店面
34 次主力店内景
35 次主力店内景

33

34

35

SUB-ANCHOR STORE
AND SPECIALTY CATERING
次主力店及特色餐饮

通州万达广场注重品牌规划，邀约了一批具有国际影响力并适合国内消费需求的优异品牌落户。比如，ZARA是西班牙INDITEX集团旗下的一个子公司，1975年设立于西班牙，是全世界名列前茅的服装品牌，在世界各地56个国家设立超过两千家的连锁店。ZARA深受全球时尚青年的喜爱，品牌优异而价格却相对低廉。引进ZARA，帮助平民实现拥抱"High Fashion"的愿望。

Focusing on the brand planning, Tongzhou Wanda Plaza has invited a group of excellent brands with international influence and catering for domestic demands. For instance, ZARA, a subsidiary corporation of INDITEX Group established in Spain in 1975, ranks top among clothing brands around the world and has opened more than 2,000 chain stores in 56 countries. Owing to its excellent brand with relatively cheap price, ZARA is deeply loved by the global fashion youth. The introduction of ZARA realizes the civilians' desire of hugging "High Fashion".

建筑简洁铝板特色

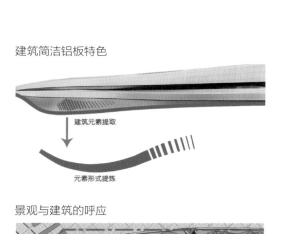

景观与建筑的呼应

39

40

41   42   43

LANDSCAPE OF PLAZA
广场景观

将景观与建筑作为一个运河场景的整体，使景观与建筑形成有机的逻辑融合。整体景观设计模拟运河场景，将建筑"比喻"成一艘船，景观模拟运河水面场景：漫江碧透，波光潋滟；轻风拂浪，海草摇曳；一艘帆船慢慢向前驶近，突然，鸢飞鱼跃，恰似帆船竞技，长风破浪，鱼水相投。

By treating the landscape and building as an entirety of canal scene, the organic and logic integration of them is created in the design. The whole landscape is designed to simulate the canal scene, of which the building is compared to a ship. The overall landscape presents the water surface scene of canal: with clear water, shimmering water surface, waving breeze and swaying seaweed, a ship slowly drifts forward, and suddenly the hawk flies and fish leap, as if sailing boats compete with each other in the river to brave the wind and the waves harmoniously.

36 广场景观全景
37 景观设计分析图
38 景观意象图
39 广场主题雕塑
40 广场主题雕塑
41 广场特色地面铺装
42 广场特色地面铺装
43 广场特色地面铺装

NIGHTSCAPE OF PLAZA
广场夜景

广场夜景设计理念是"车毂织路，舟行碧波"。作为成为一个全新的休闲与商业中心，广场采用具有流动感的铝板幕墙让人过目难忘，配合上夜景灯光的变换，犹如一艘乘风破浪的帆船。灯光暗藏于幕墙分格缝之间，既保证了夜景效果又减小了灯槽对外立面效果的破坏。

Holding the nightscape design concept of "driving on the road, boating on the river" and serving as a brand new leisure and business center, the plaza adopts the impressive aluminum curtain wall. In combination of nightscape lighting changes, the plaza is looking like a ship braving winds and waves. Concealed in diving joints of curtain walls, the lighting design both guarantees the nightscape effects and decreases the impact of light trough to façade effect.

46

47

EXTERIOR PEDESTRIAN STREET
室外步行街

室外商业步行街也延续了建筑外装折叠表皮的概念，所造出的商业走廊具有丰富的建筑表情。商业街因此出现了多样的立面变化；而这种立面的律动进而将商业空间提升为更高层次的品质体验空间。

The exterior commercial pedestrian street design inherits the concept of folding surface at building facade. The commercial corridor is created with abundant building expression. Therefore, various facade changes are found in the commercial street, while the rhythm of facade further upgrades the commercial space to a higher-level quality experience space.

44 广场外立面夜景
45 室外步行街
46 室外步行街景观小品
47 室外步行街景观小品
48 室外步行街

48

02

NANNING QINGXIU WANDA PLAZA
南宁青秀万达广场

时间 2014 / 12 / 18　**地点** 广西 / 南宁
占地面积 5.84 公顷　**建筑面积** 49.22 万平方米

OPENED ON 18th DECEMBER, 2014
LOCATION NANNING, GUANGXI ZHUANG AUTONOMOUS REGION
LAND AREA 5.84 HECTARES　**FLOOR AREA** 492,200M²

OVERVIEW OF PLAZA
广场概述

广场位于南宁市青秀区，东葛路延长线以南，贤宾路以北、滨湖路以西，地块西侧为其他规划用地，占地5.84万平方米，总建筑面积49.22万平方米。广场由购物中心、室外步行街、甲级写字楼、五星级酒店组成。其中大商业建筑面积19.57万平方米，甲级写字楼建筑面积23.88万平方米，五星级酒店建筑面积4.3万平方米。

Nanning Qingxiu Wanda Plaza is located at Qingxiu District in Nanning, and to the south of Dongge Road Extension Line, north of Xianbin Road, west of Binhu Road, with the western side being planning land. It has a land area of 58,400 square meters and the total floor area of 492,200 square meters, including 195,700 square meters for major commercial area, 238,800 square meters for Class A office building and 43,000 square meters for the five-star hotel. The plaza consists of shopping center, exterior pedestrian street, Class A office building and five-star hotel.

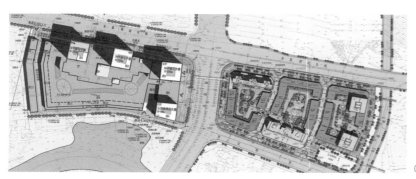

01

01 广场总平面图
02 广场全景

FACADE OF PLAZA
广场外装

南宁是一座历史悠久的古城，具有深厚的文化积淀和得天独厚的自然条件，令南宁满城皆绿，四季常青。长久以来，南宁形成了"青山环城，碧水绕城，绿树融城"的城市风格。设计提取"碧水"为元素、流动的邕江为设计形态、结合具有民族特色的"壮锦"，使建筑立面上线条纵横交错，一条动感曲线轻轻滑过，彩釉玻璃和穿孔板镶嵌其中，如同锦缎的纹理一般。这种创意设计使构图变得活泼、生动，活跃了整体商业氛围。

大商业立面用现代的设计手法，弧形线条和垂直线优雅相接，简约、时尚、轻盈，使人轻松愉悦。入口雨篷外挑，恰如空中划过的飘带，吸引人流进入商场，成为强烈的视觉中心。

Featured by profound cultural accumulation and advantaged natural conditions, Nanning is an old city with a long history, presenting an evergreen city environment. Through the ages, Nanning has emerged as "a city encircled by green hill, surrounded by blue water and filled by green trees". Extracting the "Blue Water "as the design element, the Yongjiang River as design form, and combining with "Zhuang Brocade" of national characteristics, criss-cross lines on building facade is presented with a dynamic curved line gently gliding from it and colored glazing glass and perforated plate inset within it, looking like brocade texture. This innovative design as such makes the composition lively and vivid, and creates an active business atmosphere.

The facade of large commercial area applies modern design method by elegantly connecting the arc lines and vertical lines, forming concise, fashion and light effects and creating relax and pleasure atmosphere. Awning overhanging at entrance looks like a ribbon flying across the sky, attracts pedestrian flowing into the store, and becomes a strong visual center.

04

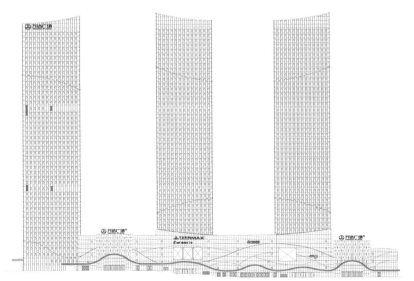

05

03 广场鸟瞰
04 广场二号入口
05 广场立面图

INTERIOR OF PLAZA
广场内装

南宁万达广场的设计概念是水。力求在内装避免过度强调装饰的手法，而是力争在空间设计上突破创新，水流弧线的元素融会贯穿整个室内步行街。圆中庭取消观光梯后，为了打破规则的圆形布局，在侧裙做外波浪造型挑板，丰富了空间。回廊宛若渡口，连桥胜似一叶扁舟，泉水在船体下轻轻围绕，托举着它飘向遥远的大海。站在一层的步行街中，抬头仰望，船体造型颇为形象；连桥打破原有直来直去的造型，而是在三维空间进行立体雕琢，创造出船的形体。

Following the design concept of water, the interior design of Nanning Wanda Plaza seeks to avoid overemphasis on decoration method, while strives to make breakthrough and innovation in spatial design. The element of flow arc runs throughout all interior pedestrian street. After the sightseeing elevator is removed from the circular atrium, in order to break the regular circular layout, the side skirt adopts the wave-shape cantilever slab, enriching the space. With ambulatory being a ferry, connecting bridge being a tiny boat, the spring water lightly flows around the boat and pushes it to drift into the distant sea. Standing on the GF of pedestrian street and looking up, the building's ship shape seems to be more vivid; breaking the original regular shape, the connecting bridge is carved in three-dimensional space to present the shape of a ship.

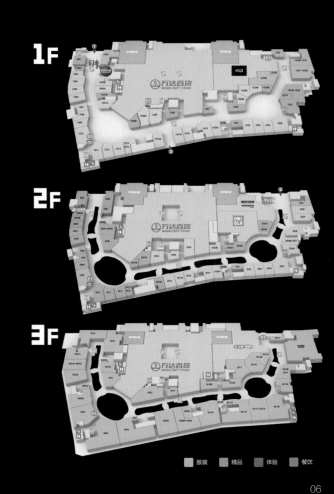

1F

2F

3F

■ 服装　■ 精品　■ 体验　■ 餐饮

06

08

09

10

11

WANDA DEPT. STORE
万达百货

南宁青秀万达百货与广场同期开业，分布于广场的一层到四层，建筑面积达3.0万平方米，从硬件设施到经营服务，均居南宁百货前列，成为改变南宁商业格局乃至成为广西人民生活和消费的中心。其中一层的主题是"珠光宝气，仙履饰界"——经营珠宝首饰、黄金手表、鞋履皮具及化妆用品；二层的主题"优雅女王，低调奢华"——经营各类女装；三层的主题是"清新俏丽，典藏时髦"，经营各类童装、少女服饰；四层的主题是"儒雅时尚，运动休闲"——经营各类男装、皮装皮具和运动休闲服饰。

Nanning Qingxiu Wanda Department Store is opened along with the Plaza and located at GF to 3F of the Plaza, covering the floor area of 30,000 square meters. Ranking the forefront among the department stores in Nanning in terms of hardware facilities and operating service, it has not only altered the business structure of Nanning but also served as the living and consumption center of Guangxi people. Themed by "Beautiful Jewelry and Fairy Accessories", GF sells jewelries, gold watches, shoes and leather goods, and cosmetics; themed by "Elegant Queen and Modest Luxury", the first floor sells women's dress; themed by "Fresh Beauty and Trendy Collection", the second floor sells children's garments and girls' clothing; themed by "Refined Fashion and Sports Leisure", the third floor sells men's clothing, leather garment and leather goods, and sportswear.

WANDA CINEMA
万达影城

南宁青秀万达影城着重突出"绿城"之四季常绿、枝繁叶茂的理念。在大堂空间内，营造了"独木成林"的意境；天花造型通过不同颜色的灯光效果，表现阳光透过树叶照射到地面的斑驳影响，给人以舒适、荫凉、安全的感觉。在重点区域，以壮族民族特色的手工制品"壮锦"作为装饰，起到点睛之用。

Nanjing Qingxiu Wanda Cinema emphasizes the philosophy of evergreen and flourish "Greentown". The lobby builds the atmosphere of "One tree making a forest"; by virtue of colorful lighting effect, the ceiling modeling gets the mottled effect that is similar to the sunshine's shining the ground through the leaves, bringing a sense of comfort, shade and safety. In key areas, the decoration of "Zhuang Brocade", the handmade articles of Zhuang ethnic characteristic, further highlights the area.

09 万达百货街景
10 万达百货内景
11 万达百货内景
12 万达影城大厅   12

13

LANDSCAPE OF PLAZA
广场景观

景观设计以壮族的传统文化为概念设计主题，着力打造一个传承地域特色文化的商业景观场所。大商业带状广场是由一系列的造型花坛与大商业建筑围合而成。花坛造型简约，结合南方特有的植物，打造出具有南宁地域特色的植物群落。雕塑将壮族岩画歌舞中人形象艺术提炼，与铜鼓结合，表现原始的张力。雕塑铜鼓的有力造型、飘逸的壮锦，尽显欢腾、欣欣向荣的景象，视觉效果简洁、大气。

The landscape design concept is themed by traditional culture of Zhuang nationality, striving to forge a commercial landscape space that inherits regional distinctive culture. The linear plaza of major commercial area is encircled by a series of shaped parterres and the major commercial buildings. The simply-shaped parterres use endemic plants in the south to build plant communities with local characteristic. Sculpture applies artistic refinement from the singing and dancing figures exhibited in rock painting of Zhuang nationality, and combines with bronze drums, presenting the original tension. The powerful modeling of sculpture and bronze drum and elegant Zhuang brocade demonstrate the jubilant and thriving scene and creates the concise and grand visual effects.

13 广场主题雕塑
14 广场花坛
15 广场景观小品
16 广场景观浮雕墙
17 广场外立面夜景

14

15

16

NIGHTSCAPE OF PLAZA
广场夜景

灯光设计风格以打造秀美绚烂的南宁城市环境形象
为目标,通过变幻多彩的灯光系统与建筑立面幕墙
的完美结合,在映衬建筑秀美的夜景体态的同时,也
提升了南宁城市的夜景景观品质。

The lighting design aims to create the graceful and gorgeous
Nanning city environmental image. Through perfectly
combining the changeable and colorful lighting system with
the building facade curtain walls, the nightscape quality of
Nanning is enhanced while the graceful nightscape shape of
buildings is displayed.

17

19

20

18 室外步行街
19 步行街景观小品
20 步行街景观小品

EXTERIOR PEDESTRIAN STREET
室外步行街

将南宁当地休闲文化，与东南亚文化巧妙融入商业
步行街，缔造出"文商并重，古今融合"的商业气息，
让南宁多元文化引领世界潮流。一条东南亚风情与
当地文化历史相结合的特色街，通过一系列的特色
小品设计，既塑造室外步行街的文化调性、贴近当地
文化、有亲切感和归属感，又便于群众记忆、形成有
趣谈资，成为旅游推广的特色，吸引外来人士。

The local leisure culture and Southeast Asian culture are
skillfully integrated into the commercial pedestrian street,
creating the commercial atmosphere of "the equal of culture
and business, the merging of the past and the present",
and making Nanning multi-culture lead the world tends. By
combining the Southeast Asia culture with the local Culture
& History and applying a series of special architecture pieces
design, the featured street not only presents cultural tone,
local culture, intimacy and sense of belonging, but also is
easy to be remembered and talked, promoting the tourism to
attract outsiders.

03

LANZHOU CHENGGUAN WANDA PLAZA
兰州城关万达广场

时间 2014 / 10 / 24　**地点** 甘肃 / 兰州
占地面积 5.5 公顷　**建筑面积** 40 万平方米

OPENED ON 24th OCTOBER, 2014
LOCATION LANZHOU, GANSU PROVINCE
LAND AREA 5.5 HECTARES　**FLOOR AREA** 400,000M²

OVERVIEW OF PLAZA
广场概述

兰州城关万达广场位于城关区天水北路景观大道旁，具有濒河和湖景双重景观优势，占地面积5.5公顷，总建筑面积40万平方米，容积率6.12，含购物中心、180米超高层甲级写字楼、超五星级酒店、超高层豪宅、室外步行街，是目前已开业万达广场中占地最小、但业态最全的城市综合体。

Lanzhou Chengguan Wanda Plaza is located beside Landscape Avenue, Tianshui North Road in Guanqu District, endowing with it a dual landscape dominance since it is close to the river and lake landscapes. It has a land area of 5.5 hectares, the total floor area of 400,000 square meters and floor area ratio of 6.12. It is the smallest but the most comprehensive city complex among Wanda Plazas, consisting of the shopping center, 180m high super high-rise Class A office building, super five-star hotel, super high-rise mansion and exterior pedestrian street.

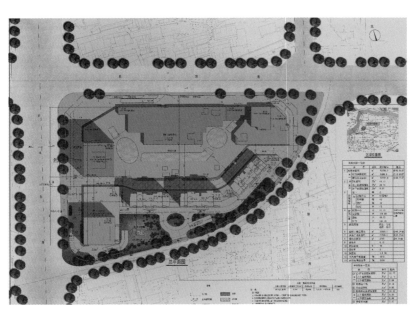

01 广场总平面图
02 广场全景

FACADE OF PLAZA
广场外装

立面设计充分发掘兰州当地的历史人文、地理、色彩等因素，将奔腾的黄河和优雅柔美的敦煌壁画作为建筑外观的缩影。色彩上，大商业以黄河色为基调，穿插丝绸的飘带香槟黄、中国红及水印象的灰白，形成主次分明、冷暖协调的建筑性格；写字楼则采取冷色调，表达其庄重，尊贵的感觉，而且也采用了与大商业外立面飘带相同的肌理，从而增强了整体的设计感。大商业的飘带设计有收有分，且层次分明；写字楼的"月牙泉"飘带外凸，酒店的飘带内凹，有呼应、有变化、有秩序，相互协调。每个室内商业街入口位置即为河流或沙漠（丘）的起伏点、制高点。根据这种关系凸显每个商业入口，加之超白玻璃后的竖向白页的衬托，犹如浪尖上的水滴，点点洒落在大地。

The facade design fully explores the local factors of humanity history, geography, color, etc., and vividly presents the epitome of the running wild Yellow River and graceful and elegant Dunhuang frescoes. In color, the large commercial area adopts the Yellow River color as primary color, interspersed with champagne yellow from silk ribbon, China red and grey impressed by water, forming the building character which is featured by clear priorities and cold-warm color harmony; the office building depends on cold color to exhibit a dignified and noble perception, and applies the ribbon texture found in the big commercial area facade to strengthen the sense of wholeness. The ribbon of big commercial area, is designed by integrating the projecting and shrinking design, presenting distinct gradation; the ribbon of "Crescent Spring" adopts convex shape in office building echoes with concave shape of hotel's ribbon, resulting in a changeable, orderly and coordinated condition. The design treats undulation point or commanding point of river or desert (dune) as entrance of each interior commercial street to highlight each entrance, and with the foil of vertical louvers behind the ultra-white glass, each entrance looks like a water drop on crest sprinkling around the ground.

04

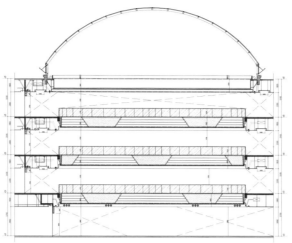

09

10

11

LANDSCAPE OF PLAZA
广场景观

"丝路飞天"是甘肃敦煌莫高窟的名片，以此为题材所设计的节点广场，展现丝绸之路上艺术与文化交融的历史场景；而"悠悠驼铃，沙漠之舟"的主题雕塑，将丝绸之路的形象用抽象的形式加以现代重塑，寓意兰州从历史走向现代的步伐。

The "Silk Road" is the name card for Dunhuang Grottoes in Gansu Province. Taking the subject as such, the node plaza design highlights the historical scene that the art integrates with the culture across the Silk Road. Moreover, themed with "Ringing Camel Bell, Ship of the Desert", the sculptures remodel the Silk Road Impression through abstract form with modern technique, which implies Lanzhou steps from the historical to the modern.

14

NIGHTSCAPE LIGHTING
夜景照明

12 景观绿化
13 景观绿化
14 广场主题雕塑
15 广场外立面夜景

15

17

18   19

16 室外步行街
17 室外步行街
18 步行街景观小品
19 步行街景观小品

EXTERIOR PEDESTRIAN STREET
室外步行街

在立面设计中，充分发掘兰州当地的地理、历史、文化、色彩等因素，并将它们转化为设计语言运用在立面设计中。
飞天、沙丘、月牙泉，都是兰州的地域符号，从中提炼"飘带"元素，运用在立面设计上，使立面飘逸灵动。商业街以
黄色、砖红色为基色，穿插灰白等辅助色，冷暖协调，主次分明，体现商业建筑的地域特色。

The facade design fully explores the local factors like geography, history, culture and color, transforming them into design language.
The design extracts "ribbon" element from the local regional symbols of Flying Apsaras, sand dunes and Crescent Moon Spring, and
by adopting the "ribbon" to facade design, creating flexible and flowing effects. The commercial street fully demonstrates the regional
characteristics of commercial building, as it adopts yellow and brick-red as primary colors and grey as adjunctive color to create
harmonious relationship between the cold color and warm colors and enjoy clear priorities.

LONGYAN WANDA PLAZA
龙岩万达广场

时间 2014 / 11 / 07　**地点** 福建 / 龙岩
占地面积 13.07 公顷　**建筑面积** 72.15 万平方米

OPENED ON 7th NOVEMBER, 2014
LOCATION LONGYAN, FUJIAN PROVINCE
LAND AREA 13.07HECTARES　**FLOOR AREA** 721,500 M²

OVERVIEW OF PLAZA
广场概述

龙岩万达广场位于龙岩市新罗区龙岩大道主轴起点，毗邻龙岩行政中心与金融中心，地理位置显赫。广场占地13.07公顷，总建筑面积72.15万平方米，其中地上部分56万平方米，地下部分16.15万平方米，包含购物中心、五星级酒店、甲级写字楼、商业街等几大功能业态。

Longyan Wanda Plaza is located at starting point of principal axis of Longyan Road, Xinluo District in Longyan and adjacent to Longyan Administration Center and Financial Centre, enjoying a prominent location. Covering a land area of 13.07 hectares and the total floor area of 721,500 square meters meters, among which 560,000 square meters is aboveground area and 161,500 square meters is underground area, the plaza encompasses shopping center, five-star hotel, Class A office building and commercial street and other functional formats.

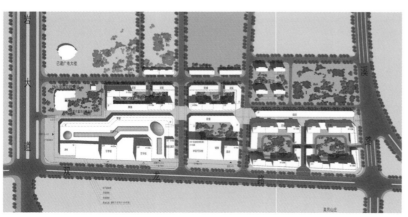

01 广场总平面图
02 广场鸟瞰

03

FACADE OF PLAZA
广场外装

建筑设计从客家土楼汲取灵感，借鉴圆形土楼层层叠落内廊，建筑外表皮形成起伏的韵律，在有限的立面空间内营造出立体感很强的层次效果，并且把整个建筑联系起来。

The architectural design sources its ideas from Hakka earth building and imitates stacked inner corridor of round earth building. Such design forms a rolling rhyme in building surface, and builds the layering effect with strong stereoscopic sensation in the limited space and effective connection of the whole building.

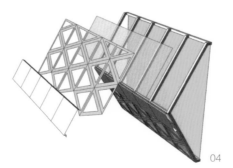

04

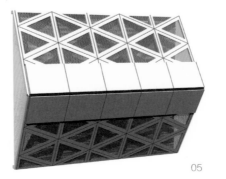

05

03 广场外立面
04 广场外立面模块 1
05 广场外立面模块 2
06 广场外立面模块 3
07 广场入口

建筑入口"方"与"圆"切合，与传统圆寨土楼形式相契合；而入口主体带有镂空图案肌理的铝板环形围绕、入口玻璃幕墙的退后，营造出入口强烈的视觉空间，使"土楼"更具有现代感。

The "square" and "round" shapes combination at the entrance boasts of harmonious coexistence with traditional round earth building. The main parts of entrance are surrounded by aluminum plate with hollow pattern texture, and the glass curtain wall at entrance is set back, building a strong sense of visual space at entrance and making the "earth building" more modern.

06

07

1F

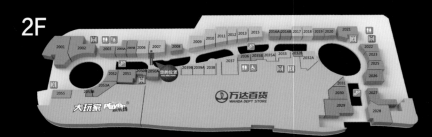

2F

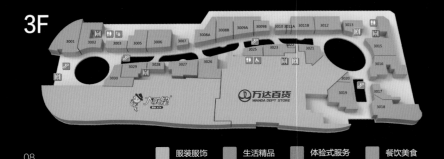

3F

■ 服装服饰	■ 生活精品	■ 体验式服务	■ 餐饮美食

INTERIOR OF PLAZA
广场内装

广场内装以土楼为出发点而展开，提取出圆形、椭圆形的建筑形态，逐渐抽象写意成流畅的曲线，并将土楼富有特色的色彩、肌理效果及外部形态，通过现代的手法巧妙地演绎出具有特色的空间效果。

Starting from the earth building, the interior design of plaza extracts round and oval building forms to gradually interpret smooth curve lines via abstraction. By adding the modern technique on the color, texture effects and interior shape of the earth building, distinctive spatial effects are skillfully interpreted.

WANDA KIDS PLACE
万达宝贝王

龙岩万达儿童天地（"万达宝贝王"）位于广场一层，建筑面积约2000平方米，是具有知识产权的创新型动漫亲子乐园。乐园定位是：孩子和家长共同参与的亲子体验空间、寓教于乐的孩子成长空间、舒适的第三休闲空间。乐园面向的目标群体为1-12岁的亲子家庭；经营理念是"让孩子在梦想中成长"；主要设施包括集合机械游乐类、攀爬城堡类、电玩类、亲子活动类、水吧和零售类六个板块，通过融入高科技元素及教育体验，打造具有动漫特色的室内亲子乐园。

The Longyan Children's Spatial ("Wanda Kids Place") is located at GF, Longyan Wanda Plaza, with the floor area of about 2,000 square meters. It is innovative cartoon parent-child paradise that has intellectual property rights. The Kids Place aims to build the Parent-child experience space where the parents and children are jointly participating; children growing up space where education is combined with recreation; and the third recreational space where parents and children enjoy comfortable perception. The kids place is opened to the family with 1 to 12 year-old children, bearing the operation philosophy is "let the children grow up in the dream". Its main facilities are divided into six areas, covering mechanical amusement facilities, castle climbing, video games, Parent-child activities, water bar and retail stores. Adding high-tech elements and educational experience, the kids place builds an indoor parent-child paradise with cartoon features.

12 万达宝贝王入口
13 万达宝贝王活动场景
14 万达宝贝王活动场景
15 万达宝贝王内景

13

14

15

LANDSCAPE OF PLAZA
广场景观

龙岩景观融合"土楼屋檐圆融层叠"的趣味形态，诠释广场"圆融凝聚"之美。景观设计着力以当地特色文化为切入点，将土楼形体之美、客家蓝印花布、客家生活细节等众多元素引用于铺地、花坛、座椅、特色水景环境细节中，给顾客提供了一个可观可赏、可驻可游、开合相间的复合景观空间。

Integrating the fun form "Harmonious and stacked Eave of Earth Building", the landscape fully interprets the plaza beauty of "Harmonious Cohesion". Proceeding from the local characteristic culture, the landscape introduces the beautiful shape of earth building, Hakkas blue calico and Hakkas living details into the design of environmental details, such as pavement, parterre, benches and characteristic waterscape, creating a retractable composite landscape space for viewing, appreciation, accommodation and traveling.

17

18

NIGHTSCAPE LIGHTING
夜景照明

铝板外表面点缀一些星光灯，呼吸式的变化节奏使人内心愉悦舒适。主立面采用铝板横向贯穿裙楼，在锥体处用LED灯洗亮凹形空腔，将建筑的"土楼"优美线条加以完美地呈现。

Several star lights embellish at the external surface of aluminum plate, of which the breathing-type rhythmic change creates inner pleasure and comfort. With aluminum plate transversely crossing the podium and LED lighting shining the concave cavity at cone, the main facade perfectly presents the graceful lines of "earth building".

19

20

21

22

19 室外步行街入口
20 步行街景观小品
21 步行街景观小品
22 步行街景观小品
23 室外步行街

EXTERIOR PEDESTRIAN STREET
室外步行街

步行街立面设计以龙岩客家土楼为灵感来源，采用方形土楼山墙尖顶与坡顶相结合的建筑形态，提炼简化土楼内部构建元素进行重新组合，木制格扇搭配米黄色墙面以现代手法重新演绎土楼印象。具有龙岩文化特色的景观小品贯穿始终，以客家人文场景雕塑点缀街道，配以阳伞坐凳、庭荫休闲组团，使整个步行街消费气氛活脱有致。

Sourcing its ideas from the Longyan Hakka earth building, the facade design adopts the building form integrating gable pinnacle of square earth building and slope crest, recombines the interior building elements refined and simplified from the earth building, and reinterprets the earth building image by mean of the modern technique of wooden lattice with beige wall. The pedestrian street is decorated with landscape articles featured by local culture throughout, interspersed with Hakkas humanities sculpture and equipped with seats under parasol and shading leisure cluster, resulting in an active and interesting consumption atmosphere.

GUANGZHOU PANYU WANDA PLAZA
广州番禺万达广场

时间 2014 / 11 / 08 **地点** 广东 / 广州
占地面积 6.75 公顷 **建筑面积** 50.396 万平方米

OPENED ON 8th NOVEMBER, 2014
LOCATION GUANGZHOU, GUANGDONG PROVINCE
LAND AREA 6.75 HECTARES **FLOOR AREA** 503, 960 M²

OVERVIEW OF PLAZA
广场概述

广州番禺万达广场位于广州市番禺区，毗邻迎宾大道、南村路及汉溪路延长线。广场占地6.75公顷，地上建筑面积35.3万平方米，是一座集甲级写字楼、SOHO、商业步行街、文化中心、休闲娱乐中心为一体的大型城市综合体。

Guangzhou Panyu Wanda Plaza is located at Panyu District in Guangzhou and close to Yingbin Avenue, Nancun Road and Hanxi Road Extension Line. The plaza occupies the land area of 6.75 hectares and the aboveground floor area of 353,000 square meters. As a large-scale city complex, it contains Class A office building, SOHO, Commercial Pedestrian Street, cultural centre and leisure centre.

01

01 广场总平面图
02 广场主入口
03 广场立面图

02

03

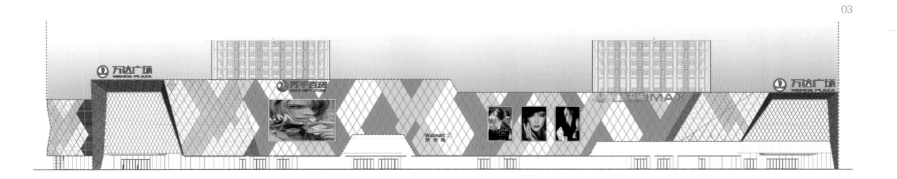

FACADE OF PLAZA
广场外装

立面设计灵感来源于奢侈品店的表皮肌理，运用构成手法，通过色彩、质感、凹凸的变化取得时尚的立面效果。立面材料主要由穿孔铝板和玻璃构成。穿孔铝板分别为白色、浅灰绿色及深灰绿色三种颜色，配色时尚，符合商业特点，结合不同的孔率、相互穿插变化、立面进退关系，在光影下形成似"鳞片"的肌理效果。

Sourcing its idea from the skin texture of luxury shops, the facade design adopts constitution technique, and through the changes of color, textures and concave-convex shapes, gets fashionable facade effects. Colored in white, light celadon and dark celadon respectively and presenting fashionable color matching, the perforated aluminum plates, filled with commercial characteristics, create the texture effects similar to scale under the light and shade in combination with various porosities, alternate variations and advance and retreat relation on facade.

04 广场侧立面
05 广场主立面

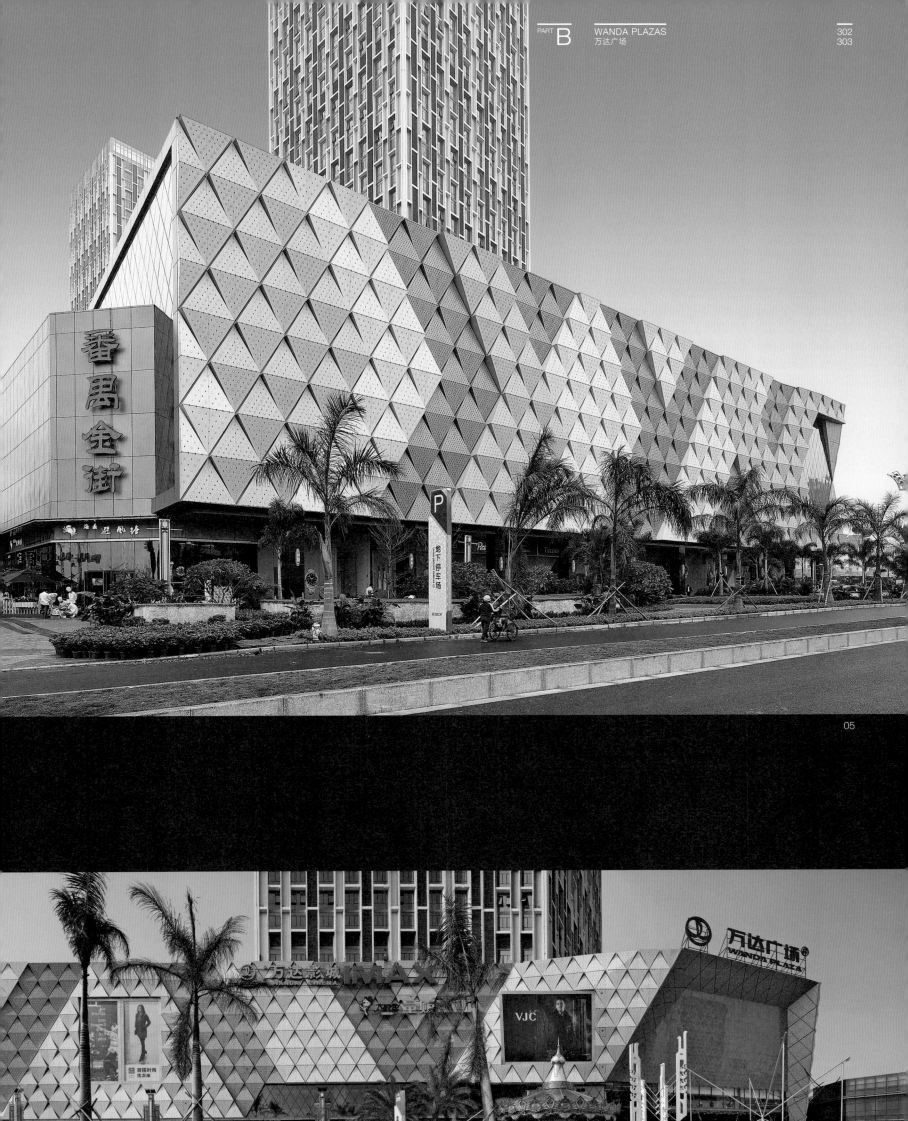

05

INTERIOR OF PLAZA
广场内装

岭南特色文化是设计表现着重点: 中山装、木棉花、岭南建筑轮廓形态等。这些岭南特色文化元素，经过抽取提炼，融进室内各空间中，形成丰富的形态和细节。

椭圆中庭空间体量大，以中山装式的观光梯与椭圆形态过桥为焦点，配合侧裙的木棉花板点状发光图案，既丰富了细节也隐喻了岭南文化。圆形中庭的空间与步行街连通，表面纹理是木棉花瓣的中庭侧裙，由此处出发，进入飘带状步行街。

The design focuses on the characteristic culture of Lingnan, such as Chinese tunic suit, Kapok, outline forms of Lingnan Architecture. By extracting and refining, characteristic culture of Lingnan is integrated into various indoor rooms, creating rich forms and details.

With huge space volume, the design of elliptical atrium focuses on Chinese tunic suit shape sightseeing elevator and ellipse shape bridge, and adds point-like light pattern on kapok board of side skirt, enriching the details and implying Lingnan culture. The circular atrium is in connection with pedestrian street,and adopts the side skirt looking like Kapok petals as surface texture. From here, the ribbon shape pedestrian street is accessible.

06 椭圆中庭
07 室内步行街
08 商铺落位图
09 室内步行街电梯

07

1F

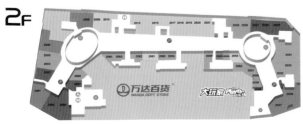

2F

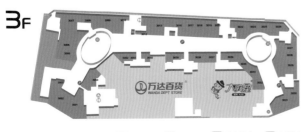

3F

服装服饰　　生活精品　　体验式服务　　餐饮美食　08

09

10 圆中庭
11 特色餐饮店铺
12 特色餐饮店铺

11

12

SUB-ANCHOR STORE AND SPECIALTY CATERING
次主力店及特色餐饮

广州番禺万达广场业态丰富，主力店有万达百货、万达影城、大歌星等；次主力店有ZARA、优衣库、GAP等适合国人需求的知名快时尚品牌。如GAP是美国最大的服装公司之一，以价格合理、式样简单为标志，深受美国大众的喜爱。广州番禺万达广场作为GAP在广州的第一家专卖店（建筑面积近1000平方米），落户羊城深受欢迎，每逢周末GAP门店都是顾客盈门。

Guangzhou Panyu Wanda Plaza is filled with various business types, of which the anchor stores cover Wanda Department Store, Wanda Cinema, the singer and so on; the sub-anchor stores include ZARA, UNIQLO, GAP and other famous and fashionable brands, catering for domestic demands. For instance, as one of the largest American clothing companies, GAP is marked by reasonable price with simple style and deeply loved by American people. As the first specialty store opened at Guangzhou Panyu Wanda Plaza (with the floor area being about 1000 square meters), GAP store wins warm praise from Guangzhou customers. It is always full of customes at weekend.

13

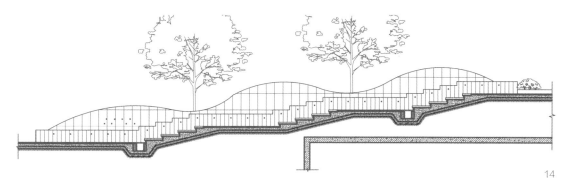

14

LANDSCAPE OF PLAZA
广场景观

景观的主题是"新岭南水乡"。借岭南传统中押韵来表达美好的意愿，分别命名东南广场为"金玉（鱼）满堂"、主力店广场为"海边拾贝"以及离不开"水"的西南水景广场，传达传统的岭南文化。

Themed with "New Lingnan Water Village", the landscape design takes advantage of traditional rhyme to express the wonderful wishes, respectively naming southeast square the "Gold and Jade (yu, a homophone for fish) Filling the Hall", the anchor store square the "Picking up Shells at Seaside" and southwest Waterscape square the "water" to convey traditional culture of Lingnan.

NIGHTSCAPE LIGHTING
夜景照明

15

16

EXTERIOR PEDESTRIAN STREET
室外步行街

室外步行街将雨篷作为立面设计的重要元素，顺应高差，此起彼伏，像纽带一样贯穿整个商铺立面；同时结合穿孔铝板的疏密变化，形成富有韵律的商业立面效果。步行街的外立面处理主要考虑人的空间尺度感受，立面处理上注重细部岭南元素的体现。

Treating the awning as an important element and going with the elevation difference, the facade design of exterior pedestrian street presents an undulating shape, looking like bond running across the overall store facade, meanwhile, in combination with density changes of perforated aluminum plates, creates a rhythmic commercial facade. It focuses on spatial dimension perception and pays attention to reflect detail elements of Lingnan.

17

YANTAI ZHIFU WANDA PLAZA
烟台芝罘万达广场

时间 2014 / 11 / 21　**地点** 山东 / 烟台
占地面积 21.09 公顷　**建筑面积** 117.96 万平方米

OPENED ON 21ST NOVEMBER, 2014
LOCATION YANTAI, SHANDONG PROVINCE
LAND AREA 21.09 HECTARES　**FLOOR AREA** 1179, 600 M²

02

OVERVIEW OF PLAZA
广场概述

广场位于烟台市芝罘区核心地带，毗邻烟台文化中心和烟台市博物馆，占据烟台中心城区的门户位置；占地21.09公顷，总建筑面积117.96万平方米，其中地上部分90.51万平方米，地下部分27.45万平方米，是集购物中心、超高层超五星级酒店、超高层甲级写字楼、城市商业街等几大功能业态为一体的大型商业综合体。

Yantan Zhifu Wanda Plaza is located at the core zone of Zhifu District in Yantai and adjacent to Yantai Cultural Center and Yantai Museum, occupying the portal position of central urban area. The plaza has a land area of 21.09 hectares and the total floor area of 1179,600 square meters, including 905,100 square meters for aboveground area and 274,500 square meters for underground area. As a large-scale commercial complex, it contains shopping center, super high-rise five-star hotel, super high-rise Class A office building and commercial street and other functional formats.

01 广场鸟瞰
02 广场立面图
03 广场总平面图

03

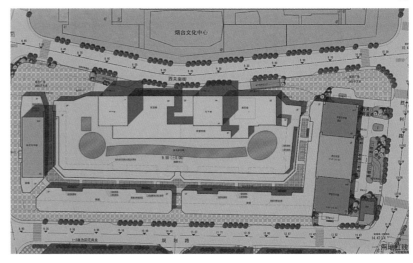

FACADE OF PLAZA
广场外装

立面设计的灵感来源于广阔的海洋和烟台的历史文化。"碧海云烟"为设计主题，通过立面白色柔软的竖向铝板肋条的起伏组合，在浪花点点的海洋蓝铝板幕墙的衬托下，形成朵朵白云从蓝蓝的海上升腾、海浪汹涌、若隐若现的效果。大商业主入口和主立面形成呼应，延续"碧海云烟"的主题，采用橙色彩釉玻璃和本体灰彩釉玻璃相间的布局。

Themed with "Sea and Clouds", the facade design is conceived from the broad ocean and the rich history and culture of Yantai. Through fluctuant combination of the soft white vertical aluminum plate rib strips, the design strives to draw a looming picture, depicting that clouds are rising from the blue sea, waves are surging against the background of the ocean blue aluminum plate curtain wall with spoondrift. Echoing with the main facade design, the main entrance of large commercial area inherits the theme of "Blue Sea with Beautiful City" by adopting alternative arrangement of orange colored glazing glass and grey colored glazing glass.

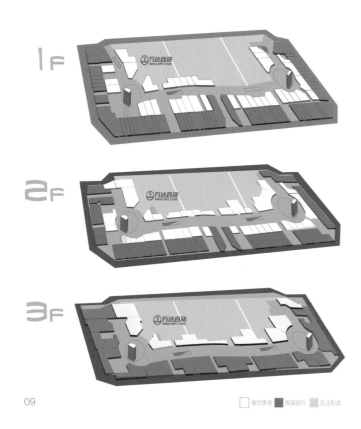

1F

2F

3F

□ 餐饮美食 ■ 服装服饰 ■ 生活配套

INTERIOR OF PLAZA
广场内装

内装以"碧海云烟"为主题，纹理板的排序方式采用渐变方式，地面石材运用流线造型进行无缝拼接，释放出海浪的概念，也寓意出对"碧海云烟"的设计理解。椭圆中庭设计运用GRG材料使其侧帮造型弧形更加圆滑，弧形造型提取于海浪涌动造型元素。

The interior design is themed with "Sea and Clouds". By applying gradient sequence to texture board, streamlined shape for seamless splice to floor stone, the design has revealed the concept of wave and implied the interpretation for the design of "Sea and Clouds". As, the elliptical atrium uses GRG material to make the lateral wall's arc shape inspired by the surging wave shape more smooth.

09 商铺落位图
10 圆中庭
11 室内步行街

12

13

14

12 万达宝贝王活动场景
13 万达宝贝王活动区
14 万达宝贝王活动区
15 万达宝贝王入口

15

WANDA KIDS PLACE
万达宝贝王

万达儿童游乐场包含淘气堡、大型机械设备、小型电玩设施等多种游乐项目。乐园分为攀爬区（淘气堡＋攀爬区）、爬山车区、海盗船区、碰碰车区等活动区域，适合不同年龄和兴趣爱好的儿童游乐。

Wanda Kids Place contains various entertainments, covering naughty castle, a large-scale mechanical equipment and small video game facilities and others, which is divided into climbing area (naughty castle+ climbing area), mountain bike area, pirate ship area, bumper car area and other activities areas, catering for children of different ages and various interests.

16

LANDSCAPE OF PLAZA
广场景观

景观设计从大商业、大交通、大广场、大文化等方面
对景观整体规划入手，以交通规划为先导，从广场与
城市的链接界面、链接节点、景观轴线入手，将广场
整体景观元素的运用与城市规划、建筑设计相结合，
形成有机统一整体。把广场设计作为景观设计的核
心，合理规划广场布局，如步行街出入口为大型集散
广场，主广场场地开阔，可以满足大型商业活动使用
要求，契合各业态定期或不定期的多种活动。

Focusing on overall planning, the landscape design is
carried out from the following aspects: large commercial
area, macro-transportation, grand plaza and macro-culture.
Guided by transportation planning and starting from the
link interface, link node between the plaza and city and
landscape axis, the design integrates the application of plaza
overall landscape elements with the urban planning and
building design, creating an organic unity. Treating the plaza
design as the core, the landscape carries out reasonable
planning on plaza layout. For instance, a large distribution
square at the entrance and exit of pedestrian street, featured
by large and widened main square, can accommodate large-
scale commercial activities as well as the various regular or
irregular activities of all kinds of business types.

17

18

16 景观主题雕塑
17 景观小品
18 景观小品
19 广场外立面夜景
20 室外步行街
21 室外步行街

NIGHTSCAPE LIGHTING
夜景照明

将大商业和写字楼的照明达到"海天一色"的整体效果加以考虑。大商业与写字楼的照明整体联动，与门头共同构成夜景主体形象；主题动画延续"碧海云烟"的一体化设计理念，以蓝色为主调体现海洋特色，大气快节奏的灯光变化体现时代脉搏；门头照明以暖黄光为主调，与蓝光形成补色对比，灯光变化相对稳重，强化了主入口的视觉焦点的标志性。

Striving to achieve the effect of "the Sea Melted into the Sky", the lighting design treats the large commercial area and office building as a whole. The overall linkage of the two and gate jointly form the main image of nightscape; the thematic animation inherits the integrated design concept of "Sea and Clouds", with blue color showing ocean characteristics and magnificent & fast-paced lighting changes reflecting the pulse of the times; the gate lighting uses the dominant warm yellow light to exhibit complementary color contrast with blue light, plus its relatively stable changing of lighting together emphasize the symbolic visual focus at entrance.

19

20

EXTERIOR PEDESTRIAN STREET
室外步行街

室外步行街通过凸凹拼图模数化的立面设计方法，将不同色彩立面单元拼接成富于变化又形式统一的立面肌理，形成商业步行街内部"一店一色"、步移景异的丰富效果。

By employing the method of modular concavo-convex puzzle, the facade design connects various facade units of different colors to build a changeable yet uniform facade texture, resulting in a rich interior effect of "One Shop, One Style" and "One Step, One Scene".

21

07

JIANGMEN
WANDA PLAZA
江门万达广场

时间 2014 / 11 / 28 **地点** 广东 / 江门
占地面积 10.56 公顷 **建筑面积** 61.52 万平方米

OPEND ON 28th NOVEMBER , 2014
LOCATION JIANGMEN, GUANGDONG PROVINCE
LAND AREA 10.56 HECTARES **FLOOR AREA** 615,200 M²

01 广场总平面图
02 广场立面图
03 广场全景

01

OVERVIEW OF PLAZA
广场概述

江门万达广场位于江门市蓬江区，紧邻江门电视台与五邑华侨广场，占地10.56公顷，地上建筑面积47.52万平方米，地下建筑面积14.00万平方米，是集高端购物中心、步行商业街、商铺、五星级酒店、甲级写字楼、公寓等几大功能业态为一体的大型城市综合体。

Jiangmen Wanda Plaza is located at Pengjiang District in Jiangmen and close to Jiangmen TV Station and Wuyi Overseas Chinese Square. It covers a land area of 10.56 hectares, and the aboveground and underground floor area of 475,200 square meters and 140,000 square meters respectively. As a large city complex, it contains the high-end shopping center, commercial pedestrian street, store, five-star hotel, Class A office building, apartment and other functional formats.

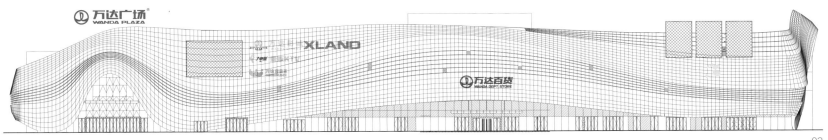

04

FACADE OF PLAZA
广场外装

江门位于珠江三角洲，当地的自然环境和侨乡文化是设计创意的源泉；丰富的水系资源和连绵起伏的山脉，是立面设计的主要元素。流动、起伏的曲线贯穿于整个商业综合体，高挑的入口着重体现"江门"的寓意。外立面造型构思灵感来源于岭南山脉的自然曲折、优美婉转的山势；行云流水的曲线，既尊重了当地历史文脉，又生动和谐地把各个单体建筑组合在一起。

Located at the Pearl River Delta, the design idea is conceived from local natural environment and home culture of overseas Chinese; the main elements of facade design sources from the rich river system and rolling hills. The flowing and rolling curve lines run across the overall commercial complex, The tall entrance focuses on demonstrating the implication of "Jianmen". Inspired by the natural, winding and beautiful Lingnan mountain range, the facade design uses freely flowing curve lines, consequently showing the reverence to local historical historical context, as well as vividly and harmoniously linking all individual buildings.

05

04 广场外立面
05 广场外立面局部

06

07

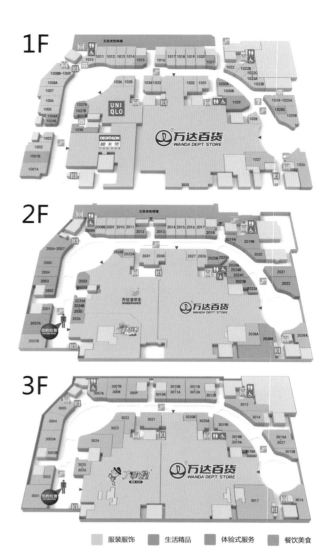

1F

2F

3F

服装服饰　生活精品　体验式服务　餐饮美食

06 椭圆中庭
07 室内步行街
08 商铺落位图
09 圆中庭

INTERIOR OF PLAZA
广场内装

椭圆采光顶穿孔铝板造型及中庭侧板飘带状图案，舞动在整个中庭空间；穿插悬台骤然张开，露出富有岭南特色的六边形肌理，成为整个空间的视觉焦点。椭圆中庭天花的穿孔铝板透出星光斑斓的效果，步行其下仿佛河流在涌动，又宛若银河般璀璨，感受到动感、绚丽的空间氛围。入口天花造型由结合地域文化特色的六边形"灯笼"组成，在灯光的映射下泛起点点星光，给人一种视觉上的享受。长街中庭侧板采用整体概念：一条曲线贯通；侧板底部天花灯带持续贯穿整个空间，产生扩张感并具有导向作用。连桥与长街侧板贯通，营造出丰富的空间感与活跃的购物气氛。

The daylighting design of the ellipse atrium adopts perforated aluminum plate modeling and ribbon-like pattern at side plates of the atriums, dancing across the whole space. The suspended platform interpersed suddenly stretches out to reveal Hexagon texture full of Lingnan characteristics, making it a visual focus. Perforated aluminium panel of the elliptical atrium ceiling creates the star-spangled effect, as if a flowing river, or the colorful galaxy, building a dynamic and colorful space atmosphere. The ceiling of the entrance is shaped by hexagon "lantern" that integrats with local cultural features, offering a visual feast under the linght. At the atriums of pedestrian street, a curve line runs across the side plates, whose ceiling light strip are found throughout the whole space, resulting in a sense of expansion and oriention. The connecting bridge connects with side plate of the pedestrian street, building a colorful space sense and active shopping atmosphere.

LANDSCAPE OF PLAZA
广场景观

景观以回忆江门的"山水情"为轴、"侨乡情"为脉，通过打造场景、营造空间、刻画细节来描述一幅"山水江门、文化侨乡"的美景。大商业以"山水侨乡"为主题，地面铺装采用像书法一样优美流畅的曲线，表达江门山水的理念。

The landscape design aims to recall the Jiangmen memory, combining the axes of "Feelings of Mountains and Waters" with the context of "Feeling of Hometown". By creating scene, building space and depicting details to present a beatuful picture of "Picturesque Jiangmen, Cultural Hometown". Taking "Scenery Hometown" as the theme, the pavement of large commercial area adopts curved lines looking like graceful and smooth calligraphy, expressing the concept of Jiangmen scenery.

11

12

10

NIGHTSCAPE LIGHTING
夜景照明

10 广场景观
11 景观绿化带
12 景观主题雕塑
13 广场夜景
14 室外步行街

13

EXTERIOR PEDESTRIAN STREET
室外步行街

设计主题是"一江山水，侨乡深情"，以此表达历史的沉淀、文化的延续。内街飘过一股岭南特色风，把岭南印象简约地表现在外街各个店铺之上。

Themed with "A River Landscape, A Deep Homesickness", the exterior pedestrian street design strives to express historical sediment and cultural continuity. All stores have thus concisely manifested the Lingnan impression, with Lingnan style wind blowing across interior street.

14

FUQING
WANDA PLAZA
福清万达广场

时间 2014 / 12 / 05　**地点** 福建 / 福清
占地面积 8.0 公顷　**建筑面积** 24.52 万平方米

OPEND ON 5ᵗʰ DECEMBER , 2014
LOCATION FUQING, FUJIAN PROVINCE
LAND AREA 8.0 HECTARES　**FLOOR AREA** 245,200M²

OVERVIEW OF PLAZA
广场概述

福清万达广场位于福建省福清市音西街道，占地8.0公顷，总建筑面积24.52万平方米，含室外步行街、购物中心、甲级写字楼等业态。

Located at Yinxi Street, Fuqing city in Fujian and occupying a land of 8.0 hectares and total floor area of 245,200 square meters, Fuqing Wanda Plaza consists of exterior Pedestrian Street, shopping center, Class A office building. etc.

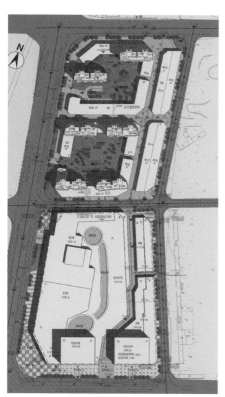

01 广场总平面图
02 广场鸟瞰

03

04

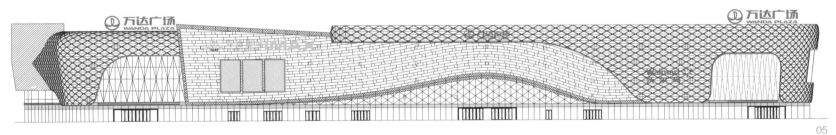

05

FACADE OF PLAZA
广场外装

建筑主体立面宛若游龙，整体造型浑然一体。体块之间相互交错映衬，好似游龙的鳞片；而色彩上的统一性，更强化了这种整体感觉。百货的入口好似游龙激起的浪花，自由平滑的曲线顺势而起，与建筑形态衔接自然。办公塔楼设计暗合了福清悠久的佛教文化，汲取了传统寺塔建筑挺拔的形体和层层檐口的肌理，创造出富有传统韵味的双塔建筑形象。

The main facade is designed in a perfectly integrated shape, seeming to be a wandering dragon. The staggered blocks, looking like the dragon scales, achieves the color harmony, highlighting its overall feeling. Shaped like the waves stirred by the dragon, the entrances at department store adopts the smooth curved lines in natural connection with plaza form. The office tower design coincides with long-standing Buddhist culture in Fuqing, drawing on tall and straight shapes and layered eave textures from traditional temple and tower, aiming to build a traditional twin tower image.

03 广场正立面
04 广场侧立面
05 广场正立面图

INTERIOR OF PLAZA
广场内装

福清是美丽的滨海城市，故设计以"海洋"为题，赋予空间生命力。设计通过动感流线和虚化元素的运用，把海洋生物的造型融入其中。椭圆形中庭贝壳图案冲孔铝板的侧帮是海洋元素的延伸，地面拼花仿佛是柔软的沙滩。黄色的LED灯带，如贝壳里暗藏珍珠的隐隐光芒。圆形中庭的地面拼花延续黄金沙滩的元素，采光顶让"撒"下来的光线更加柔和；侧帮上时而稀疏时而密集的鱼群，使空间充满流动性。

Fuqing is a beautiful coastal city, so the plaza design adopts the theme of "Ocean", endowing a vivid vitality to the space. By means of applying dynamic streamlines and imaginary elements in design, the marine organism modeling is vividly demonstrated. The perforated aluminum plate with shell pattern is used in the lateral wall of ellipse atrium to symbolize the extension of ocean. Besides, the parquet floor imitates the soft sand beach. The LED lamps flicker yellow light, looking like the faint light from concealed pearl in shell. The parquet floor of the circular atrium is consistent with elements of golden sand beach, and daylighting roof makes the sprinkled light gentler. The fish pattern is unevenly arranged at the lateral wall of atrium, making the space filled with mobility.

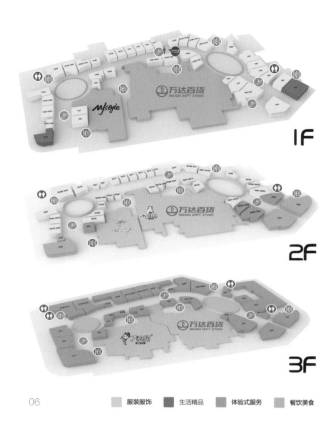

1F

2F

3F

06　　服装服饰　生活精品　体验式服务　餐饮美食

08

06 商铺落位图
07 圆中庭
08 室内步行街

LANDSCAPE OF PLAZA
广场景观

景观设计灵感缘于福清优美的自然风光和当地的特色鱼文化。主景雕塑以"渔趣"为文化脉络——鱼群的布置在平面布局上结合地面铺装的肌理走向，形成"S"形的变化；从而形成鱼群仿佛从远处向大门聚集的感受。

The landscape design sources its idea from beautiful natural scenery and the local fish culture. The main sculptures adopt the "Fun of Fishing" as culture context. In combination with the direction of pavement texture, the fish is arranged to form the S-shape variation, presenting a gathering feeling that the fish swim from afar to the gate.

10

NIGHTSCAPE OF PLAZA
广场夜景

11

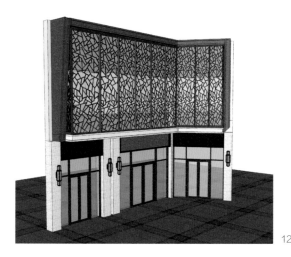

12

EXTERIOR PEDESTRIAN STREET
室外步行街

室外步行街将传统与现代的元素综合，古典的木棂窗和现代的彩釉玻璃幕墙在这里碰撞，新的铝板材料用传统的砖墙砌筑方法形成肌理。在这里人们能够忆起往昔，也可以憧憬未来，见证这座古老而又年轻的城市的发展。

The design of exterior pedestrian street integrates the traditional and modern elements, where classic wooden lattice windows collides with modern colored glazing glass curtain wall and where the modern aluminum material combines with traditional brick wall laying method to form the texture. Recalling the past and planning the future, people may witness the development of the ancient yet young city standing on the street.

12 步行街立面分析图
13 室外步行街
14 室外步行街

14

WENZHOU PINGYANG WANDA PLAZA
温州平阳万达广场

时间 2014 / 12 / 06 **地点** 浙江 / 温州
占地面积 11.93 公顷 **建筑面积** 55 万平方米

OPEND ON 6th DECEMBER, 2014
LOCATION WENZHOU, ZHEJIANG PROVINCE
LAND AREA 11.93 HECTARES **FLOOR AREA** 550,000M²

OVERVIEW OF PLAZA
广场概述

广场位于温州市平阳县鳌江镇滨江区，地势较为平坦，基本呈规则的矩形；占地11.93公顷，总建筑面积55.00万平方米。广场集购物、休闲、娱乐、餐饮、居住于一体，涵盖超市、百货、家电、电玩、KTV、影城及时尚精品街、潮流服饰街、餐饮美食街等各业态。在未来的发展中，万达广场将成为滨江中央商务区乃至鳌江两岸的商业地标。

Wenzhou Pingyang Wanda Plaza is located at Binjiang District, Aojiang Town, Pingyang County in Wenzhou, with relatively flat terrain and basically in rectangle shape. It covers a land area of 11.93 hectares and the total floor area of 550,000 square meters. The plaza integrates the functions of shopping, leisure, recreation, food & beverage and residence, covering the commercial formats of supermarket, department store, household appliances, video games, KTV, cinema, fashion and boutique street, fashion cloths street and food street. Wenzhou Pingyang Wanda Plaza is planned to develop into the commercial landmark of Binjiang central business district and of Aojiang River.

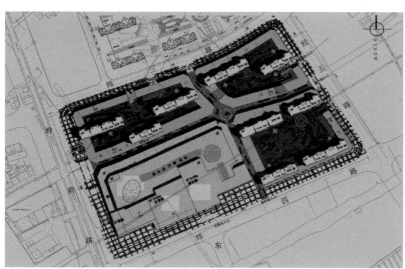

01

01 广场总平面图
02 广场正立面
03 广场侧立面
04 广场主入口

02

03   04

05

FACADE OF PLAZA
广场外装

幕墙外立面整体设计，吸取了鳌江传统文化中"龙"的元素。南立面下部，采用玻璃幕墙体系，结合大商业广场形成"鳌江"的江面；而南立面上部，采用白色的铝板幕墙体系，通过穿插组合形成"龙"的曲折多变形象。弯曲的"龙头"、"龙尾"被设计为大商业的两个主要入口。整个南侧外立面寓意"龙游鳌江"民间习俗。建筑入口顺应外立面形体走向设计，整体感强，明朗大气；印刷玻璃的材质凸显建筑的时尚与高贵。铝板幕墙采用冰裂纹（瓷器上的冰裂纹，或称"开片"）作为基本元素的表皮肌理处理手法，通过参数化的设计，力求使用冰裂纹作为元素塑造细部复杂、整体统一协调的外观形象。

The overall design of curtain wall facade is inspired from "dragon" element in traditional Aojiang River culture. The bottom of south facade uses the glass curtain system and combines with the square in large commercial area to form the shape of "Aojiang River" surface, while its top adopts the white aluminum plate curtain wall system and depicts a bending "dragon" through intersection and combination. The two main entrances of the large commercial area are designed in the shape of dragon's head and tail. Consequently, the overall south facade implies the folk custom, namely "Swimming Dragon in Aojiang River". The design of entrances echoes with facade direction, resulting in a strong, bright and grand sense; meanwhile, the glass printing material used emphasizes the fashionable and noble architectural style. In surface texture treatment, with the basic element of cracked ice crackle (or gracked glaze) coupled with parameterization design, the aluminum curtain wall strives to build an appearance featured by complex details and unified whole.

05 广场外立面
06 广场二号入口

06

INTERIOR OF PLAZA
广场内装

椭圆中庭侧帮采用了整体空间的处理手法，通过线条的勾勒以及象征江流的曲线纹样装饰的挑台，结合地面叠块拼花新颖的处理手法，使整个空间更加整体而生动。室内步行街的设计主题是以建筑本体简练的几何形体为原型，结合"折纸游戏"进行演变；整个空间中装点着无数的延伸折面，它们犹如打开的折纸，在空间中无处不在。

The lateral wall of the elliptical atrium design is in consistence with the overall spatial treatment technique, using the lines and cantilever platform decorated with curved patterns symbolizing the river. Such design combines with the novel parquet floor, making the overall space more integrated and vivid. Treating the concise geometrical form from building noumenon as prototype, the design theme of interior pedestrian street is developing from the "Paper Folding". In this manner, the whole space is penetrated by numerous extended folding surfaces, like the opened folding papers floating everywhere.

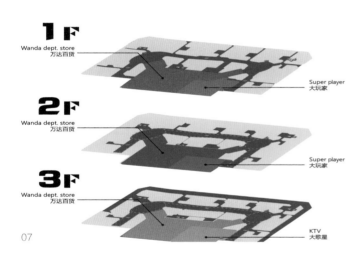

1F
Wanda dept. store
万达百货

Super player
大玩家

2F
Wanda dept. store
万达百货

Super player
大玩家

3F
Wanda dept. store
万达百货

KTV
大歌星

07

07 商铺落位图
08 室内步行街
09 圆中庭裙板
10 圆中庭

08

09

1

LANDSCAPE OF PLAZA
广场景观

景观设计以"瓯海瓷韵"（鸥瓷色青白，开缀冰裂纹）为主题，紧紧围绕建筑的功能、形态、景观，对冰裂纹元素进行提炼与再创造，很好地诠释了建筑与景观的一体化设计。主雕塑形象圆润壮观，金色表面，上覆殴瓷冰裂纹，寓意两晋、唐、宋等时期，瓯瓷不仅在国内受到欢迎，还吸引了大量国外商人前来订购，为如今温州商海圆融通达埋下伏笔。

Themed with "Porcelain Charm in Ouhai" (bluish white Ouhai Porcelain, cracked ice crackle) , the landscape is refining and recreating cracked ice crackle elements closely based on the building function, form and landscape, well interpreting the integrated design of building and landscape. The main sculpture is designed in rounded and grandeur shape, with golden surface covered with cracked ice crackle. It implies the popularity of Ouhai Porcelain at home and abroad, attracting a large number of foreign orders in the two Jin, Tang and Song dynasties, which paves the way for today's thriving business in Wenzhou.

2

13

EXTERIOR PEDESTRIAN STREET
室外步行街

商业街强调"一店一色"、移步易景，利用建筑风格
的混搭和错位布置，并有效地配合景观铺地、小品、
绿植及美陈店招设计，最大限度地提升购物体验。

Focusing on "One Shop, One Style" and "One Step,
One Scene" and utilizing mixture style and displacement
arrangement, the design effectively arranges landscape
pavement, featured landscape, green plants and display
design to fully improve shopping experience.

14

11 景观主题雕塑
12 景观主题雕塑
13 广场夜景
14 室外步行街入口
15 室外步行街

15

HANGZHOU GONGSHU
WANDA PLAZA
杭州拱墅万达广场

时间 2014 / 12 / 12　**地点** 浙江 / 杭州
占地面积 8.34 公顷　**建筑面积** 36.02 万平方米

OPEND ON 12th DECEMBER , 2014
LOCATION HANGZHOU, ZHEJIANG PROVINCE
LAND AREA 8.34 HECTARES　**FLOOR AREA** 360,200M[2]

01 广场外立面
02 广场总平面图

01

OVERVIEW OF PLAZA
广场概述

杭州拱墅万达广场位于杭州市拱墅区，占地8.34公顷，总建筑面积36.02万平方米（其中大商业9.0万平方米，写字楼11.0万平方米，室外步行街5.32万平方米，地下建筑面积10.70万平方米）。

Hangzhou Gongshu Wanda Plaza is located at Gongshu District in Hangzhou, occupying a land area of 8.34 hectares and total floor area of 360,200 square meters (including 90,000 square meters for large commercial area, 110,000 square meters for office building, 53,200 square meters for exterior pedestrian street, 107,000 square meters for underground floor area).

02

03

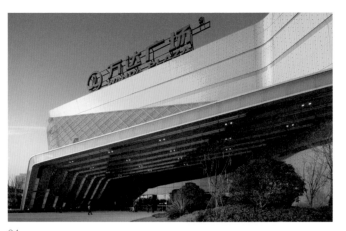

04

05

FACADE OF PLAZA
广场外装

建筑外表皮借鉴了丝绸织物特性的精髓——光滑、闪亮、流动,在阳光的照耀和微风的拂动下尽显流光溢彩。立面材质结合穿孔铝板、金属网、玻璃幕墙的组合运用,使广场的巨大体量包围在如丝绸般轻柔的表皮之下,既如丝如水、清逸流动,又坚如磐石、巍然屹立,实乃"外柔内刚"在建筑上的完美诠释。主入口立面层次感十足,无论凸显的大雨篷还是雨篷上方的金属网及穿孔铝板,均与入口大门布置交相呼应。

To be colorful and beautiful under the sunshine and blowing breeze, the plaza surface design is drawing on the essence of silk fabric—smooth, shiny and flowing. Through combinative utilization of the perforated aluminum plate, metal net and glass curtain wall, the facade design strives to make the huge plaza encircled by the silky soft surface, presenting gentle and flowing yet rock-solid and rock-firm appearance. The design perfectly interprets the concept of "Soft Outside and Hard Inside". The main entrance is full of layering and realizes a coordinated arrangement between obvious large canopy, metal net and perforated aluminum plate above the canopy with entrance gate.

06

03 广场外立面特写
04 广场入口
05 广场入口近景
06 广场入口雨篷

07

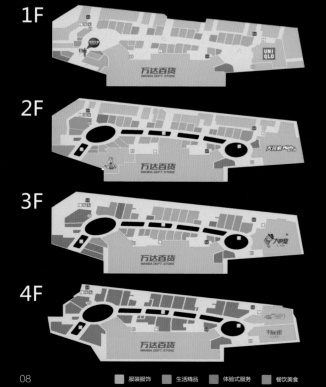

INTERIOR OF PLAZA
广场内装

广场内装的意念是"飞羽"，力求在空间设计上突破创新，契合"飞羽"概念中羽毛元素的轻盈，使流线融会贯通地穿行于整个室内步行街。圆形中庭在建筑上增加了自动扶梯，通过造型与自动扶梯的结合，使整个空间连成一体。圆形地面造型同样是直街的延续，与椭圆地面组成一个蜿蜒流转的直街、水流图案，营造出了高端、时尚的繁华商业氛围。"飞羽"的设计理念在这里得以升华。

Based on the idea of "flight feather" and for the sake of achieving the breakthrough innovations on space design, the interior design of plaza adopts the lithe feature of feather, making the streamline run across the overall interior pedestrian street. The escalator is added at the circular atrium and uses the modeling combination to link the overall space into an organic whole. As a continuation of straight gallery, the circular floor modeling combines with elliptical floor to create a winding straight gallery with flow pattern, resulting in a high-end and fashionable commercial atmosphere and sublimating the concept of "flight feather" simultaneously.

服装服饰　生活精品　体验式服务　餐饮美食

WANDA DEPT. STORE
万达百货

杭州拱墅万达百货位于广场的地上1~4层，建筑面积约2.0万平方米；定位于城市副中心旗舰店，辐射全市，动线规划和室内设计达到国际水准。杭州拱墅万达百货浓缩了当下国际、国内的时尚名品，其中一层是"都会名品馆"——经营黄金珠宝、化妆品、男女鞋及钟表；二层是"淑女亲子馆"——经营各类淑女装及童装童品；三层是"少女内衣馆"，经营各类内衣、少女服饰；四层是"绅士运动馆"——经营各类男装及运动休闲服饰。

Hangzhou Gongshu Wanda Department Store is located at GF to 3F of the Plaza, covering a floor area about 20,000 square meters. It aims at flagship store in sub-center and influence touching the whole city, and has international-level Traffic Flow Planning and Interior Design. The department store encompasses current internationally and domestically famous and fashionable brands. Themed by "Famous Urban Brand", the GF sells jewelries, cosmetics, shoes and watches; themed by "Ladies and Kids", the first floor sells various lady's dress, children's garments and products; themed by "Young Ladies Underwear", the second floor sells underwear and girls' clothing; themed with "Gentleman Sportswear", the third floor sells all kinds of men's clothing and sportswear.

10 万达百货店面
11 室内步行街商铺内景
12 室内步行街商铺内景
13 广场水景
14 景观绿化

10

11

12

PART B   WANDA PLAZAS
万达广场   352
353

13

LANDSCAPE OF PLAZA
广场景观

14

景观设计创意上体现了"寻西湖、印杭州"的概念，汲取水的形态，运用水晕、水流的流线，应用于铺装。通过砖、石材、洗米石的变幻结合，将各个景观节点有机地统一起来，以做到景观各要素的完美结合，形成统一的整体。大商业广场花坛设计再现"西湖印象"主题概念，将砚台的形态运用于花坛，使种植池雕塑化，同时强化景观空间的围合感，营造令人舒适的空间氛围。

Following the concept of "Seeking the West Lake, Reflecting Hangzhou", The design creativity of landscape extracts water forms and then applies the streamlines of water wave and water flow to the pavement. By virtue of the variational combination of bricks, stone and granitic plaster, the design organically integrates various landscape nodes, forming an integrated whole with various elements being perfectly matched. Representing the theme of "West Lake Impression", the parterre design of large commercial area adopts the ink-stone shape by sculpturing planting pool and simultaneously strengthening the sense of closure, thus creating a comfortable atmosphere.

NIGHTSCAPE OF PLAZA
广场夜景

夜景照明将极具地方文化特色的丝绸作为设计元素，建筑裙楼勾勒的线条灯犹如随风飘扬的丝带，通过灯光的变化渲染出建筑的灵动及江南的柔美风韵。立面采用三种不同的建筑肌理形成几个大的体块互相穿插。

The nightlighting design takes the silk with local cultural characteristics as the design element. The podium buildings draw the outline with lamps, looking like the flying silk ribbon. And through lighting changes, such design renders the spirituality of building and highlights graceful Jiangnan (lower reaches of the Changjiang River) flavor. The several staggered large blocks in three different textures are found in facade design.

17

18

19

20

21

EXTERIOR PEDESTRIAN STREET
室外步行街

设计以"水墨山水画"为主题，借山水画的智慧与精髓——墨为核心，境界为灵魂。反映在设计中——墨的是山、蓝的是水，山水相依相伴；层叠的屋檐宛如山，飘逸的玻璃犹如水，山水穿梭。系列主题雕塑、灯具，结合江南风情的建筑外装，将西湖梦幻美景展示得淋漓尽致。无论是再现的"曲院风荷"、"花港观鱼"、"平湖秋月"、"三潭影月"，还是书卷似的内街灯，再或是水晕般的铺装，皆让在这里休闲购物的人们仿若处于梦境之中。

Themed with "Chinese Landscape Painting", the design borrows ink, the wisdom and essence of landscape painting, as the core and artistic conception as the soul. The design uses ink color for mountain and blue color for water to represent the harmony of the two; it shapes the layered eave as mountain and flowing glass as water to show an intersectional picture of the two. The design vividly demonstrates fantastic and beautiful West Lake scenery by adopting a series of theme sculptures, lamps and integrating building facade with Jiangnan flavor. The reappeared landscape of "Lotus Stirred by Breeze in Quyuan Garden", "History of Viewing Fish at Flower Pond", "Autumn Moon over the Calm Lake" and "Three Pools Mirroring the Moon", the scroll shape interior street lamps and water wave like pavement create a "dreamland" atmosphere for customers.

YINCHUAN XIXIA WANDA PLAZA
银川西夏万达广场

时间 2014 / 07 / 18　**地点** 宁夏 / 银川
占地面积 12.0公顷　**建筑面积** 62.5万平方米

OPENED ON 18ᵗʰ JULY, 2014
LOCATION YINCHUAN, NINGXIA HUI AUTONOMOUS REGION
LAND AREA 12.0 HECTARES　**FLOOR AREA** 625,000M²

OVERVIEW OF PLAZA
广场概述

银川西夏万达广场毗邻贺兰山体育馆和宁夏大学，与金波湖相望，自然环境优越；占地12.0公顷，总建筑面积62.5万平方米，是一个包含大型购物中心、步行街、公寓和住宅等多功能于一体的大型城市综合体。

Close to Helanshan Stadium and Ningxia University and faces Jinbo Lake, Yinchuan Xixia Wanda Plaza boasts of superior natural environment. The plaza covers a land area of 12.0 hectares and the total floor area of 625,000 square meters. It is a large city complex integrating such functions as large shopping center, pedestrian street, apartment and residence.

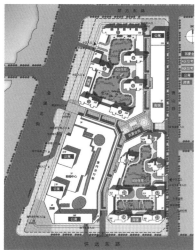

01

02

01 广场总平面图
02 广场全景
03 广场外立面

FACADE OF PLAZA
广场外装

商业大立面立意造型取自宁夏贺兰山阙和贺兰宝石，加上富有民族特色的尖券入口，形成独特的银川"万达广场"；流畅而富有韵律的横向线条，既有山脉层层叠加之意，又有黄河奔腾而去之势。

The large commercial area facade design is based on Helanshan Mountain Gap and Helan Gemstone and in combination with pointed arch entrance with national features, thus building a distinct Yinchuan Wanda plaza; the design also uses the smooth and rhythmic horizontal lines, presenting the images of overlaying mountain and gesture of rushing Yellow River.

INTERIOR OF PLAZA
广场内装

广场室内设计的特点可归纳为：洁白、纯净、崇高、圆润。椭圆中庭设计源于起伏的沙漠，又不限于沙漠的形式，色调统一、气场恢宏，给人以强烈的视觉震撼。圆中庭以优雅的弧线设计，GRG材料的细密精致的图案配合出独特的造型设计，如翩翩起舞的裙摆，放荡不羁的沙漠；又如川流不息的河水，给人无限的遐想……

The interior design features can be summarized to be white, pure, lofty and round. The elliptical atrium is originated from rolling desert, but not limited by its form. Moreover, coupled with the unified color and grand momentum, the interior design of plaza creates an intense visual impact. The design of circular atrium adopts elegant arc and uses GRG material with fine and delicate patterns for building a unique shape, looking like dancing skirt, bohemian desert and never-ending river water, and arousing endless imagination…

05

04 室内步行街
05 圆中庭裙板
06 室内步行街

06

07

08

LANDSCAPE OF PLAZA
广场景观

广场景观设计中提炼了"凤凰"这个代表银川古城历史文化的图腾作为设计元素，形成了景观概念——"凤舞九天，祥耀万达"。中心广场中特色造型拼花和主题雕塑，展现出"凤凰涅槃，浴火新生"的深刻内涵。向四个方向延展的步行街，如凤凰张开的五彩羽翼，使步行空间的界面更加生动多姿。

The landscape design forms the landscape concept –"Phoenix Dance in the Sky, Auspiciousness Blessing on Wanda", by refining the design element of "Phoenix", a totem representing history and culture in ancient Yinchuan. The central plaza is designed with distinctive splice modeling and theme sculptures, adequately explaining the profound meaning of "Phoenix Nirvana, a New Lease of Life". The pedestrian street extends to four directions as if colorful spreading wings, making the pedestrian space interface more picturesque and colorful.

NIGHTSCAPE OF PLAZA
广场夜景

夜景照明重点塑造了购物中心以及SOHO沿街立面，将两个大商业入口以及万达百货入口采用LED点阵形成一个显示屏，展现银川当地的文化特色，并突出了入口效果。

The nightscape lighting emphasizes on building facade along the street of shopping center and SOHO. The entrances of two large commercial areas and Wanda department store are equipped with the LED display screen for the sake of showing the local culture and highlighting entrance position.

09

EXTERIOR PEDESTRIAN STREET
室外步行街

西夏金街作为购物中心的延续，通过强化中部广场及各出入口的开放性、指示性，使其成为为联系内外步行街的主要过渡区域。设计抓住步行街作为三度空间的特点，把街道地面和两旁建筑物墙面所围成的相对封闭的空间加以整体打造，形成浓厚的商业氛围。

07 景观主题雕塑
08 景观小品
09 广场夜景
10 室外步行街

As the extension of shopping center, the Xixia Golden Street is served as a major transition region by highlighting the open and directive functions of central plaza and all entrances and exits. Taking notice of characteristics of pedestrian street as a three-dimensional space, the design treats the relatively closed space surrounded by street ground and building walls on both sides as a whole, building a flourishing commercial atmosphere.

10

12

MANZHOULI
WANDA PLAZA
满洲里万达广场

时间 2014 / 06 / 27　**地点** 内蒙古 / 满洲里
占地面积 8.14 公顷　**建筑面积** 10.35 万平方米

OPEND ON 27th JUNE, 2014
LOCATION MANZHOULI, INNER MONGOLIA AUTONOMOUS REGION
LAND AREA 8.14 HECTARES　**FLOOR AREA** 103,500M²

01

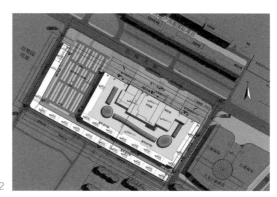

01 广场主入口
02 广场总平面图
03 广场鸟瞰

OVERVIEW OF PLAZA
广场概述

满洲里万达广场位于满洲里市互贸区，占地8.14公顷，总建筑面积10.35万平方米，包含万达百货、大玩家、大歌星、万达影城、步行街、超市等多种业态，是集餐饮、娱乐、购物、观光旅游、住宿为一体的多功能、高档次、智能化的大型综合性广场。

Manzhouli Wanda Plaza is located at Humao District in Manzhouli, occupying a land area of 8.14 hectares and total floor area of 103,500 square meters. Covering Wanda Department Store, Super Player, Super Star KTV, Wanda Cinema, pedestrian street, supermarket and other functional formats, it forms a multi-functional, high-grade and intelligent large-scale comprehensive plaza integrating food & beverage, recreation, shopping, sightseeing tour and accommodation.

03

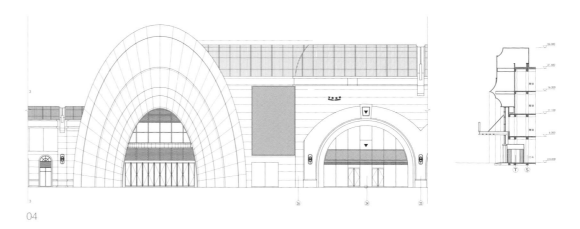

04

FACADE OF PLAZA
广场外装

建筑立面选用与俄罗斯红场建筑风格相似的砖红色材质，顶部的圆弧形装饰面具有蒙古包的神韵，使广场在展现万达文化的同时，体现满洲里"东亚之窗"的独特地域文化。大商业门头是在草原蒙古包的原型基础上，通过曲线条为主的构图元素重构而成，为满洲里浓厚的地方气息增添了新亮点。

The facade design selects the brick-red material similar to Russia's Red Square architectural style. The circular arc decorative surface at the top of plaza directly embodies the Mongolian yurt's charm. Such design of plaza demonstrates the Wanda culture, as well as the unique regional culture of Manzhouli as "Window of the East Asia". Based on the prototype of Mongolian yurt, the door of large commercial area is rebuilt by mainly using curve lines, adding the new highlight for the strong local flavor.

05

06

INTERIOR OF PLAZA
广场内装

为了让广场更具吸引力，设计从满洲里的地理环境入手，选用叶子的造型并抽象化，但所用材质并非绿色。室内步行街的主入口、地面、天花，均围绕"叶子"的主题元素展开，整体上协调、呼应。

To make plaza more attractive, the design starts from the geographical environment of Manzhouli by selecting and abstracting the shape of leaves without using green materials. The main entrance, floor and ceiling design of the interior pedestrian street are closely around the theme of "Leaf", resulting in a harmonious and congruous effect on the whole.

07

08

09

LANDSCAPE OF PLAZA
广场景观

满洲里原称"霍勒津布拉格"蒙语意为"旺盛的泉水"。在整个景观设计中，无论小品、铺装或者种植池，无不诠释泉水的脉络、泉水的韵律、泉水所激起的朵朵浪花。

Manzhouli was formerly known as "Horace Jin Prague", which means " Vigorous Spring" in Mongolian language. In the whole landscape design, the featured landscape, pavement and planting pool all strive to interpret the context, rhythm and splashing waves of spring water.

10

11

NIGHTSCAPE OF PLAZA
广场夜景

大商业夜景照明以符合当地主色调的暖光为主，夜景采用地埋灯的方式把墙面打亮，结合各立柱的壁灯及LED小功率洗墙灯形成层次感。大商业橱窗采用内透方式与三个主入口玻璃幕墙空腔相呼应，显示出空间感及立体感。

The nightscape lighting of the large commercial area adopts the warm light conforming to local dominant color. The underground lamp used to lighten the wall space and wall lamps of various columns and small power LED wash wall lamps together form a sense of layering. The show windows of the large commercial area adopt the method of internal transparency to echo with glass curtain wall cavity of three main entrances, delivering the spaciousness and stereoscopic sensation.

12

EXTERIOR PEDESTRIAN STREET
室外步行街

室外步行街整体采用地域风格，通过整合不同的颜色体块，体现出室外街外在的丰富性与活泼性，以及满洲里独特的城市文化。通过体块穿插的处理，结合满洲里地域文化对于外立面的要求，在建筑立面上形成个性化的风格。

The overall design of exterior pedestrian street exhibits the regional style; and by integrating various colorful blocks, the design demonstrates the rich and energetic external characteristic as well as unique urban culture in Manzhouli. Adopting the cross block technique and catering for the local cultural requirements on facade, the facade is built with an individuality.

13

WANDA
HOTELS
万达酒店

WANDA
COMMERCIAL
PLANNING
2014

武汉万达瑞华酒店
南宁万达文华酒店
兰州万达文华酒店
马鞍山万达嘉华酒店
荆州万达嘉华酒店
龙岩万达嘉华酒店
江门万达嘉华酒店
芜湖万达嘉华酒店
蚌埠万达嘉华酒店

01

WANDA REIGN WUHAN
武汉万达瑞华酒店

时间 2014 / 03 / 29　**地点** 湖北 / 武汉
建筑面积 7.08 万平方米

OPENED ON 29ᵗʰ MARCH, 2014
LOCATION WUHAN, HUBEI PROVINCE
FLOOR AREA 70,800M²

OVERVIEW OF HOTEL
酒店概况

武汉万达瑞华酒店位于武汉市武昌区东湖路，作为万达首个顶级奢华酒店，酒店之奢华程度可与世界任意一家顶级酒店比肩。独具特色的"钻石"立面造型与"红灯笼"秀场相映生辉，并一举拿下中国旅游业界"2014年度最佳新开业奢华酒店奖"。

Located on the East Lake Road, Wuchang District, Wuhan, Wanda Reign Wuhan, as the first seven-star hotel of Wanda Group, is comparable to any top hotel all around the world in terms of luxury. Unique "diamond" facade shape of the Hotel echoes with the "Red Lantern" show theatre, promoting the Hotel to be "The Best New Luxury Hotel " in 2014 in chinese tourism industry.

01 酒店外立面

02 酒店外立面

FACADE OF HOTEL
酒店外装

瑞华酒店的外立面设计由英国建筑事务所Make Architects完成。酒店紧邻一座象征中国传统红灯笼形态的秀场，设计师意在通过立面设计，强调这两座建筑的现代棱角和传统圆润的视觉差别，以对比突显两者间活跃的律动。

酒店的立面由902组六角形模块单元组成，让客房的窗户化身为一双双"眼睛"，俯瞰城市美景。每一块六角形模块由反光度较高的铝板组成，且以一定的角度倾出立面，在构成幕墙富有凹凸变化的整体肌理的同时，也为每个房间提供遮阳效果。六角形模块内暗藏了LED灯，让此凹凸变换的肌理效果在夜间大放异彩。

Facade design of Wanda Reign Wuhan is completed by Make Architects, a British architectural firm. In view that the Hotel is adjacent to a show theatre in the shape of Chinese traditional red lantern, designers aim to strengthen visual difference and highlight contrast of vivid rhythm between modern angular shape and traditional sleek shape through facade design.

Facade of the Hotel is composed of 902 sets of hexagonal modular units, making windows of rooms to be eyes enjoying beautiful scenery of the city. Each hexagonal module is composed of aluminum plates with high reflectance in a certain inclined angle, not only forming a rugged texture of the curtain wall, and also serving as sunshade for each room. In addition, LED light is set in the module in a concealed manner to light up the rugged texture as night falls.

04

05

INTERIOR OF HOTEL
酒店内装

06

03 酒店远景
04 酒店外立面特写
05 酒店外立面特写
06 酒店大堂

LANDSCAPE OF HOTEL
酒店景观

景观设计延续建筑立面的造型，入口水景的菱形水钵选用拉丝面不锈钢材质，结合水下灯的照射，产生晶莹剔透、交相辉映的尊贵感，同时水钵在竖向上由四周向中间升起，错落有致，形成"香槟叠水"的视觉效果。酒店入口两侧水景首次采用奢侈品店的幕墙工艺，悬浮在水景上方的抽象鱼跃造型，更是暗合东湖美景和"愉悦"的谐音。

Landscape design of the Hotel follows the building facade style, diamond-shaped fountain feature at the entrance adopts brushed stainless steel bowl to create a crystal clear sense of nobility with underwater lights. Moreover, the bowl fountain is vertically raised up in the middle from all corners, forming a visual effect of "champagne cascading". Water features on both sides of the hotel entrance firstly adopt curtain wall process for luxury stores and abstract diving fish design floating above the water features coincides with the beautiful scenery of East Lake and meaning of "pleasure" whose pronunciation is same as that of the diving fish in Chinese.

07

08

09

10

NIGHTSCAPE OF HOTEL
酒店夜景

11

WANDA VISTA NANNING
南宁万达文华酒店

时间 2014 / 12 / 18　**地点** 广西 / 南宁
建筑面积 4.95 万平方米

OPEND ON 18th DECEMBER, 2014
LOCATION NANNING, GUANGXI ZHUANG AUTONOMOUS REGION
FLOOR AREA 49,500M²

OVERVIEW OF HOTEL
酒店概况

南宁万达文华酒店位于南宁市青秀区东葛路延长线，总建筑面积4.95万平方米，地上24层，客房333间。首层为900平方米的挑空大堂、660平方米的全日餐厅、190平方米的大堂吧及340平方米的特色餐厅；二层为会议区，共设大小会议室共5间；三层设有200平方米会见厅、新娘房和1400平方米的宴会厅；四层是酒店的康体设施，含游泳池、美容美发、跳操、健身等功能；五层为中餐厅，设大中小中餐包间共8间及中餐散座区；塔楼为客房层，含标准客房、套房、部长套房、总统套房等客房类型，塔楼最顶部两层为酒店会所。

Wanda Vista Nanning is located at Dongge Road Extension Line, Qingxiu District in Nanning. It occupies a total floor area of 49,500 square meters and has 24 floors above ground and 333 guestrooms. GF consists of a higher level hall of 900 square meters, a 660 square meters All Day Dining Restaurant, a 190 square meters Lobby Bar, and a 340 square meters Specialty Restaurant; the first floor is set with meeting area, with five meeting rooms in difference sizes; the second floor is set with a 200 square meters presence chamber, bride room and a 1400 square meters banquet hall; the third floor is for fitness and recreation, including swimming pool, cosmetology & hairdressing, aerobics, fitness and other functions; the fourth floor is for Chinese restaurant with eight big, medium and small private rooms and extra seats area; the tower is for guestrooms, which include Standard Rooms, Junior Suites, Ministerial Suites, and Presidential Suites and Hotel Club at top two floors.

01

FACADE OF HOTEL
酒店外装

"万紫千红引蜂来"，酒店立面灯光通过一些灯具的有机组合，在立面形成"蜂蝶"造型，裙楼利用均匀的轮廓效果组合成起伏的花丛，正如南宁城市"四季花开"的壮美景色。

Following the concept of "Blooming Flowers Attracting Bee", the design organically combines the lamps at hotel façade to form a shape of "Bee and Butterfly"; and utilizes the podium buildings' even outline to create the effect of up-and-down flowers, resembling the magnificent and beautiful picture of "Four Seasons Flowers" in Nanning city.

03 酒店入口近景
04 酒店大堂

INTERIOR OF HOTEL
酒店内装

04

LANDSCAPE OF HOTEL
酒店景观

以壮族文化元素为设计符号，在水景、雕塑小品以及灯具的设计上着重体现特色。
整体酒店的设计不论是铺装还是植物，都在细节上精细布置，使其精致富有格调。

Accepting the Zhuang cultural elements as the design symbols, the design of waterscape,
sculptures and lamps are also focusing on local characteristics. The overall design of hotel
emphasizes fine arrangement on details to create exquisite style, whether on pavements or plants.

05 酒店景观
06 景观小品特写
07 景观小品特写
08 酒店景观雕塑
09 酒店夜景

08

NIGHTSCAPE OF HOTEL
酒店夜景

09

03

WANDA VISTA LANZHOU
兰州万达文华酒店

时间 2014 / 10 / 24　**地点** 甘肃 / 兰州
建筑面积 4.17 万平方米

OPEND ON 24th OCTOBER, 2014
LOCATION LANZHOU, GANSU PROVINCE
FLOOR AREA 41,700M²

01

OVERVIEW OF HOTEL
酒店概况

兰州万达文华酒店位于兰州市城关区天水北路，毗邻
兰州水车博览园景区。酒店建筑面积4.17万平方米，地
下2层，地上18层，设有各种客房307间；配套设施齐
全，包含宴会厅、会议室、全日制餐厅、中餐厅、特色餐
厅、健身中心、美容美发、行政酒廊和地下停车场等。

Wanda Vista Lanzhou is located at Taishui North Road,
Chengguan District in Lanzhou and close to Lanzhou
Waterwheel Garden. The hotel occupies a floor area of
41,700 square meters, with 2-storey and 18-storey under
and above the ground repspectively. It has 307 guestrooms
of various kinds and well-equipped supporting facilities,
encompassing banquet hall, meeting room, all day dining
restaurant, Chinese restaurant, specialty restaurant, fitness
center, cosmetology & hairdressing, Executive Lounge and
underground parking , etc.

01 酒店设计手绘稿
02 酒店远景
03 酒店近景

02

03

万达文华酒店 WandaVist

04

FACADE OF HOTEL
酒店外装

酒店与写字楼、大商业的立面肌理相呼应：舞动的飘带造型内凹，有呼应、有变化、有潜在的秩序，整体设计感强。设计通过与兰州印象、地域特色、文化元素的结合，着重表达了特点鲜明、形式丰富的立面设计意图。

酒店入口是汇集目光、增加气势，给人强烈印象的第一道关口。出挑的14米、宽28米长的古铜色雨篷，以简洁的凹凸横向收边描绘出精致的外形。天花造型以双层玻璃及黄色透光云石的穿插为主要结构，白天如璀璨水晶，夜间配合灯光，晶莹煽动、高贵典雅。

The hotel is echoing with office building and large commercial area in aspect of facade texture: the concave shaped dancing ribbons have consistency, variation, potential order and a strong sense of design. Through combination with Lanzhou impression, regional characteristics and cultural elements, the design strives to express the intent of building a facade with distinct features and rich forms.

As the eye-catching, powerful and impressive perception is firstly felt at hotel entrance, it adopts the 28m wide bronze canopy cantilevered for 14m, and concise, concave-convex horizontal margin lines to sketch the delicate appearance. The ceiling adopts the double glazing glass interspersed with yellow translucent marble as the main structure, looking like the gleaming crystals at the daytime and forming an elegant and noble atmosphere with crystal glittering under the light at night.

INTERIOR OF HOTEL
酒店内装

05

NIGHTSCAPE OF HOTEL
酒店夜景

04 酒店入口
05 酒店大堂
06 酒店夜景

WANDA REALM MA'ANSHAN
马鞍山万达嘉华酒店

时间 2014 / 09 / 19 **地点** 安徽 / 马鞍山
建筑面积 3.6 万平方米

OPENED ON 19ᵗʰ SEPTEMBER, 2014
LOCATION MA'ANSHAN, ANHUI PROVINCE
FLOOR AREA 36,000M²

01 酒店外立面
02 酒店入口

01

OVERVIEW OF HOTEL
酒店概况

马鞍山万达嘉华酒店位于安徽省马鞍山市太白大道和东湖路交口处，总建筑面积3.6万平方米，拥有客房285套。酒店主体建筑地上17层，建筑面积3万平方米；地下2层，建筑面积0.6万平方米。酒店拥有相应的配套设施，包括：首层大堂面积750平方米，挑空层高13米，首层还设置大堂吧及全日餐厅等；二层为中餐厅及特色餐厅；三层为宴会、会议层，包括一个面积为1200平方米的大宴会厅及多个大小会议室；四层为康体层，包括健身房、游泳池等康体设施；客房楼顶部的行政楼层包含高级套房、行政酒廊以及总统套房。

Wanda Realm Ma'anshan is located in the intersection of Taibai Avenue and Donghu Road, with total floor area of 36,000 square meters and 285 guest rooms. As for its main building, there are totally 17 overground floors with floor area of 30,000 square meters and 2 underground floors with floor area of 6,000 square meters. Its supporting facilities are listed by floor. The ground floor accommodates a 75 square meters lobby, 13m high raised layer, lobby bar and all day dinning restaurant etc. The first floor is used for Chinese restaurant and specialty restaurant. The second floor serves banquet and conference, including a huge banquet hall with area of 1,200 square meters and several large and small meeting rooms. The third floor is for fitness and recreation, including gym, swimming pool, etc. The administrative floor on top of guest room building contains deluxe suites, executive lounges, and presidential suites.

03

FACADE OF HOTEL
酒店外装

外立面设计中将灵动且富有生机的"绿叶"作为创作语言，抽象并提取其精华，将其叶脉及纹理造型重新整合赋予建筑立面。在向自然表达敬意的同时将建筑与周边景色相呼应，使建筑造型得以扩展延续，并使立面处理上以流畅的叶子轮廓曲线划分虚实外墙材料。在整体玻璃幕墙外，以白色铝板为主材的纹理装饰作为外立面的主要设计语言。上下贯通的铝材造型会根据不同位置的需要或加宽或变窄，抽象地构成了绿叶及叶脉造型。

Taking vibrant and vivid "green leaf" as the creation language, the facade design abstract and draw essence of it and reintegrate leaf vein and texture shape. While expressing respect to the nature, the design harmonizes the building with surrounding landscape to extend the building shape, and distinguishes the real and virtual external wall materials with fluent profile curve of leaf when handling facade. Beyond the whole glass curtain wall, white aluminum plate acting as principal material of the texture decoration is adopted as the principal design language. Widening or narrowing of aluminum product running from top to bottom can be done in accordance with different location demand, which abstractly constitutes the shape of leaf and vein.

03 酒店入口
04 酒店大堂

INTERIOR OF HOTEL
酒店内装

04

05

06

LANDSCAPE OF HOTEL
酒店景观

景观设计整体按照功能标准设置VIP停车、大巴停车、旗台，特设水景依然贯穿沿用整个项目整体的"花山叶雨"为理念。酒店前主体水景设计用"山"语言，充分表达出马鞍"山"特色。酒店侧墙水景"九山环一湖"的形式形成山峦起伏的气势。地面细节为市花"桂花"的金属片，体现了马鞍山嘉华酒店的尊贵与文化。

In accordance with functional standards, the landscape design sets up VIP parking, bus parking, flag platform, and apply the project overall concept of "flower mountain and leaf rain" to special waterscape. "Mountain" language is used for the waterscape design of main part in front of hotel, fully expressing the "mountain" character of Ma'anshan. The "Nine Mountains Surrounding a Lake" form of hotel flank waterscape forms mountainous momentum. Sheetmetal of city flower "Osmanthus Fragrant" constitutes the floor details, showing dignity and culture of the hotel.

07

08

05 酒店入口喷泉
06 酒店入口喷泉特写
07 酒店景观
08 酒店景观
09 酒店夜景

NIGHTSCAPE OF HOTEL
酒店夜景

09

WANDA REALM JINGZHOU
荆州万达嘉华酒店

时间 2014 / 09 / 20 **地点** 湖北 / 荆州
建筑面积 3.69 万平方米

OPENED ON 20th SEPTEMBER, 2014
LOCATION JINGZHOU, HUBEI PROVINCE
FLOOR AREA 36,900M²

01 酒店外立面
02 酒店鸟瞰

OVERVIEW OF HOTEL
酒店概况

荆州万达嘉华酒店位于北京西路，总建筑面积3.69万平方米，其中：主体建筑地上17层，建筑面积3.09万平方米；地下2层，建筑面积0.6万平方米。酒店设客房285套，拥有五星级酒店相应的配套设施，包括：酒店首层大堂面积800平方米，挑空层高13米，首层还设置大堂吧、全日餐厅及特色餐厅等；二层为中餐厅；三层为宴会、会议层，包括一个面积为1200平方米的大宴会厅及多个大小会议室；四层为康体层，包括健身房、游泳池等康体设施；客房楼顶部的行政楼层包含高级套房、行政酒廊以及总统套房。

Wanda Realm Jingzhou is located at Beijing West Road with total floor area of 36,900 square meters. As for its main building of hotel, there are totally 17 overground floors with floor area of 30,900 square meters and 2 underground floors with floor area of 6,000 square meters. The hotel totally has 285 guest rooms and supporting facilities required by five-star hotel, The ground floor accommodates a 800 square meters lobby, 13m high raised layer, lobby bar, all day dinning restaurant, specialty restaurant etc. The first floor is used for Chinese restaurant. The second floor serves banquet and conference, including a huge banquet hall with area of 1,200 square meters and several large and small meeting rooms. The third floor is for fitness and recreation, including gym, swimming pool, etc. The administrative floor on top of guest room building contains deluxe suites, executive lounges, and presidential suites.

FACADE OF HOTEL
酒店外装

酒店立面立意"竿和流水"，富于现代感的时代气息，硬朗的竖向线条也起到遮阳作用，符合绿色建筑标准。同时，竖向铝板幕墙的折板造型，也体现出节节向上的强烈势态。主体与裙房顶部采用收缩处理，更加使得建筑极具活力，同时也增强了建筑空间层次感。

主入口简洁大方，金属铝材、高透玻璃及深色石材交相辉映，厚重而不失灵动，磅礴气势油然而生。雨篷外围的折板造型与主题建筑语言相呼应，充分从细节处延续了整体的建筑语言。

Following the concept of "Rod and Running Water", the hotel facade is rich in modern atmosphere. Hard vertical lines not only have shading effect, but also meet standards of green building. At the same time, the folded plate modeling of vertical aluminum sheet curtain wall also reflects the strong upward momentum. The contraction treatment of main part and podium top makes building more dynamic, and enhances layering sense of building space as well.

Main entrance is simple and generous. Aluminum product, high transparent glass and dark stone material add radiance and beauty to each other, presenting dignified yet flexible and majestic momentum. Folded plate model of canopy periphery echoes with architectural thematic language, extending overall architectural language in details.

05

INTERIOR OF HOTEL
酒店内装

06

07

LANDSCAPE OF HOTEL
酒店景观

酒店后庭院打造成一个自然怡人的花园，休闲的木平台依水而建。在水面远处坐落着一个休憩凉亭，景亭前面是跌水，背面是茂密的树丛，营造一处世外桃源。

Creating the hotel courtyard into a natural and pleasant garden, with leisure wooden platform built beside the water. A rest pavilion stands in the distance, with its front facing water drop and back dense thickets. All these create a fictitious land of peace.

08

NIGHTSCAPE OF HOTEL
酒店夜景

09

WANDA REALM
LONGYAN
龙岩万达嘉华酒店

时间 2014 / 11 / 07　**地点** 福建 / 龙岩
建筑面积 3.83 平方米

OPENED ON 7ᵗʰ NOVEMBER, 2014
LOCATION LONGYAN, FUJIAN PROVINCE
FLOOR AREA 38,300M²

01 酒店远景
02 酒店近景

OVERVIEW OF HOTEL
酒店概况

龙岩万达嘉华酒店位于龙岩市新罗区双龙路，共拥有305间典雅舒适的客房。酒店拥有汇聚东方美食精髓的"品珍"中餐厅；囊括全球美味的"美食汇"全日餐厅以及极具现代时尚气息的"焱"高档特色扒房；呈现各式茶茗、环球珍藏佳酿的大堂酒廊。酒店拥有1200平方米的无柱式大宴会厅，以及配备先进设施的会议厅。

Wanda Realm Longyan is located on the Shuanglong Road, Xinluo District, Longyan City, having 305 comfortable and elegant guest rooms in total. The hotel has "Zhen" Chinese Restaurant gathering oriental cuisine essence, all day "Café Vista" converging delicacy worldwide, high-grade "Yan" special steak house with modern elements, lobby lounge serving all kinds of tea and globally treasured wine, 1,200 square meters non-column grand banquet hall and well-equipped conference hall.

01

03 酒店入口
04 酒店大堂

FACADE OF HOTEL
酒店外装

酒店设计灵感来自于土楼外墙上斑驳交错的龟裂纹，并且让这种错位纹理自下而上逐渐变化，呼应土楼墙体所具有的下实上虚的特点。设计强调垂直空间的层次感，营造出一种仿佛置身于东方古城堡的舒畅感受。立面通过每隔三层进行一次竖向铝板构件的粗细变化，体现土楼上间隔出现的小窗，使立面在统一的竖向感中拥有细节的变化。在裙房处增加了玻璃幕墙体块的穿插，既增加了大堂和休息厅的通透性，也增强了建筑形体的虚实关系和肌理对比。

Hotel design is inspired by the mottled and interlaced turtle crack on the external wall of earth building, and makes the dislocation texture change gradually from bottom to top, echoing with the characteristics of earth building wall with upper virtuality and lower reality. Emphasizing level sense of vertical space, the design tries to create a feeling as comfortable as staying in eastern castle. Through thickness change of vertical aluminum components every three layers, the facade highlights the spaced small windows of earth building, making the facade varied in details in the vertical sense of unity. Alternately increasing of glass curtain wall blocks on the podium not only enhances permeability of lobby and lounge, but also strengthens virtual-real relationship and texture contrast of building shape.

INTERIOR OF HOTEL
酒店内装

LANDSCAPE OF HOTEL
酒店景观

酒店景观展示了浓郁的地方特色风貌，秉承"圆融凝聚之美"的理念，将当地土楼、梯田提炼和升华，融入酒店后庭廊架、梯级的弯曲水台之中，形成微缩梯田、地方风貌景观，既体现了客家民风民俗又提升了酒店的文化品位。

To show strong local characteristics and by adhering to the concept of "Harmonious Cohesion", the hotel landscape design refines and sublimates local earth building and terrace, and then integrates them into winding water platform of backyard corridor gallery frame and steps, forming miniature terrace and local landscape. The design as such not only reflects the Hakka folk customs, but also enhances the hotel's cultural taste.

05

07

05 景观亭
06 酒店水景
07 酒店水景
08 酒店夜景

NIGHTSCAPE OF HOTEL
酒店夜景

08

07

WANDA REALM JIANGMEN
江门万达嘉华酒店

时间 2014 / 11 / 28　地点 广东 / 江门
建筑面积 4.15 万平方米

OPENED ON 28th NOVEMBER, 2014
LOCATION JIANGMEN, GUANGDONG PROVINCE
FLOOR AREA 41,500M²

02

OVERVIEW OF HOTEL
酒店概况

江门万达嘉华酒店位于江门市发展大道，总建筑面积4.15万平方米，其中：地上建筑面积3.5万平方米；地下建筑面积0.65万平方米。酒店设客房325套，拥有五星级酒店相应的配套设施，包括：酒店首层大堂面积约760平方米，挑空层高13米，首层还设置大堂吧、全日餐厅及特色餐厅等；二层为中餐厅；三层为宴会、会议层；四层为康体层，包括健身房、游泳池等康体设施；顶部行政楼层包含高级套房、行政酒廊以及总统套房。

Wanda Realm Jiangmen is located at Fazhan Road of Jiangmen, with total floor area of 41,500 square meters, including 35,000 square meters and 6,500 square meters for overground and underground area respectively. The hotel totally has 325 guest rooms and supporting facilities required by five-star hotel, The ground floor accommodates a 760 square meters lobby, 13m high raised layer, lobby bar, all day dinning restaurant, specialty restaurant etc. The first floor is used for Chinese restaurant. The second floor serves banquet and conference. The third floor is for fitness and recreation, including gym, swimming pool, etc.; the administrative floor on top of guest room building contains deluxe suites, executive lounges, and presidential suites.

FACADE OF HOTEL
酒店外装

酒店作为江门万达广场一个组成部分，在立面设计中呼应大商业的水系飘带，采用仿生手法，似微风吹过，水面涟漪飘荡，回归于自然。富丽堂皇不再是酒店的专有词，还兼有自然生态，使得酒店整体悠然而生，宁静高雅。

As a part of Jiangmen Wanda Plaza, the hotel echoes with water ribbon of large commercial area in its design, adopting bionic method to resemble the natural perception of soft breeze and drifting water ripples. In this manner, the hotel, in addition to the inborn gorgeousness, is endowed with natural ecology, making the hotel being carefree, quiet and elegant.

01 酒店外立面
02 酒店入口

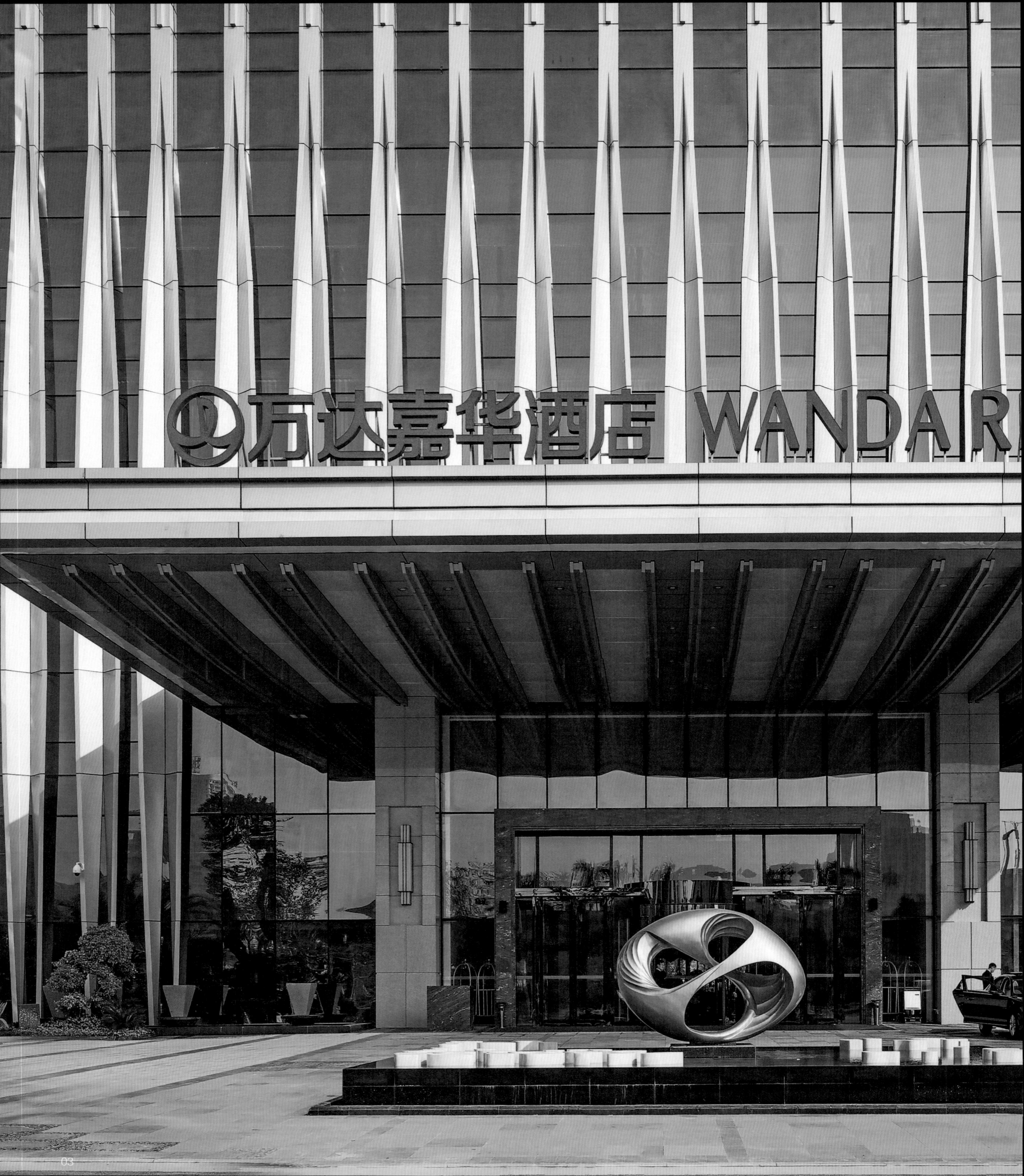
万达嘉华酒店 WANDA R

03 酒店入口
04 酒店大堂

INTERIOR OF HOTEL
酒店内装

05

LANDSCAPE OF HOTEL
酒店景观

用曲线变化来勾勒出大空间与趣味的小空间，以古朴简洁的铺装和酒店建筑相呼应，小品雕塑等通过对江门市花"三角梅"花型元素的提取，与种植设计、灯光效果交相辉映，既营造出传统的岭南风格，又能在此基础上辅以现代的景观，传承现代与传统的完美统一。

Using curve change to sketch a large space and the small interesting space, traditional and simple decoration echoes with hotel building, and the element of bougainvillea, city flower of Jiangmen, in small sculptures etc. to add radiance to each other with planting design and lighting effects, traditional Lingnan-style is delivered and perfect unity of modern and traditional elements are available in combination with modern landscape.

06

07

NIGHTSCAPE OF HOTEL
酒店夜景

08

WANDA REALM WUHU
芜湖万达嘉华酒店

时间 2014 / 12 / 06　**地点** 安徽 / 芜湖
建筑面积 3.71 万平方米

OPENED ON 6th DECEMBER, 2014
LOCATION WUHU, ANHUI PROVINCE
FLOOR AREA 37,100M²

01 酒店远景
02 酒店近景

01

OVERVIEW OF HOTEL
酒店概况

芜湖万达嘉华酒店坐落于芜湖市北京中路，总面积
3.71万平方米，地上层数18层，客房283间。首层为挑
空大堂、全日餐厅、大堂吧和特色餐厅；二层为中餐
厅；三层设有会见厅、会议室和宴会厅；四层是酒店的
康体设施，含游泳池、健身跳操、美容美发等功能。塔
楼为客房层，含标准客房、套房、部长套房、总统套房
等客房类型，17层为行政酒廊。

Wanda Realm Wuhu is located in Beijingzhong Road, Wuhu
City, with total floor area of 37,100 square meters. It has
18 aboveground floors and 283 guest rooms. The ground
floor accommodates a high lobby, lobby bar, all day dinning
restaurant, specialty restaurant etc. The first floor is used
for Chinese restaurant. The second floor serves presence
chamber, banquet hall and meeting room. The third floor
is for fitness and recreation, including gym, swimming
pool, cosmetology and hairdressing, etc.; the tower is for
guestrooms, which include Standard Rooms, Junior Suites,
Ministerial Suites, and Presidential Suites and executive
lounge at 17F.

FACADE OF HOTEL
酒店外装

外立面设计充分考虑临水的优势地理位置，通过建筑立面层层律动的波浪造型，作为建筑统一的主题和肌理，与水面相互辉映，形成协调统一并充满趣味性的建筑形象。酒店犹如漂浮在水面的帆船，轻盈明亮，倒映于湖水中。酒店的裙房设计灵感来源于竹子，与酒店塔楼的横向线条及有序的排列形成鲜明的对比。

Taking full account of the advantageous waterfront location, the facade design employs the layered wave shape as unified theme and texture of building to be merged with water, forming coherent and unified building full of fun. The hotel is shaped like a sailing boat that is floating on the water and reflected in the lake, being light and bright. The design of podium is inspired by the bamboo, standing in stark contrast with the horizontal lines and orderly arrangement of tower building.

03 酒店外立面
04 酒店外立面特写
05 酒店外立面特写
06 酒店外立面特写
07 酒店大堂

04

05

06

**INTERIOR
OF HOTEL
酒店内装**

07

LANDSCAPE OF HOTEL
酒店景观

景观设计以当地的"徽派"文化为背景，结合当地的现状，以现代设计手法来渲染和营造安徽本土文化的精神和文化氛围。酒店前场通过提取"徽派建筑文化"中的元素，采用现代手法加以运用，将"徽派元素"渗入到构筑物、雕塑、铺装以及整体空间布置中。整个前场以水滴状"叠水"与水滴"雕塑"作为景观的视觉中心，既保证了视线的通透又起到一定的虚影作用，与中国园林的"通而不透"的设计手法形成契合。景墙周围应用中式的瓦片镂空和中式花钵，起到烘托景墙的作用，同时进一步强化了景观的现代与古韵结合的风格。

In the context of local "Anhui-style" culture and combination with local situation, the landscape design renders and creates spiritual and cultural atmosphere of Anhui local culture. By extracting elements from "Anhui-style architectural culture", the front of house applies modern techniques to make sure Anhui style elements be enrolled into structures, sculpture, pavement and the integral space layout. The whole front of house adopts waterdrop-shaped "stacking water" and water drop "sculpture" as the visual center of the landscape, which not only ensures the unobstruction of sight, but also plays a role in creating virtual image. Such design method echoes with that of "unobstructed yet opaque" in Chinese garden. Hollowed-out Chinese tile and flowerpot applied on landscape wall highlights the wall, and at the same time, strengthens the landscape combination style of old and new.

08 酒店花园
09 石桥
10 景观灯
11 酒店夜景

09

10

NIGHTSCAPE OF HOTEL
酒店夜景

11

WANDA REALM
BENGBU
蚌埠万达嘉华酒店

时间 2014 / 12 / 12　**地点** 安徽 / 蚌埠
建筑面积 3.44 万平方米

OPENED ON 12th DECEMBER, 2014
LOCATION BENGBU, ANHUI PROVINCE
FLOOR AREA 34,400M²

OVERVIEW OF HOTEL
酒店概况

蚌埠万达嘉华酒店位于蚌埠市蚌山区东海大道，主体地上建筑共17层，拥有285间舒适豪华的客房和套房，拥有9米高的1000平方米无柱式超大宴会厅，并配有健身中心、室内恒温游泳池和行政酒廊等设施。

Wanda Realm Bengbu is located in Donghai Avenue, Bengshan District, Bengbu City. As for its main building, there are totally 17 overground floors, 285 comfortable and luxurious rooms and suites. A 1000 square meters pillar-free super large banquet hall with height of 9m and supporting facilities, such as a fitness center, indoor heated swimming pool and an executive lounge, etc.

01 酒店入口
02 酒店外立面

万达嘉华酒店 WANDA REALM

03

03 酒店入口
04 酒店大堂

FACADE OF HOTEL
酒店外装

建筑设计汲取了蚌埠自然景观中的"淮河流水"和地方文化中的"折扇"意向。酒店平面采用轻微转折的方式，既保证了面积平衡，也可以得到近似折扇的形象。立面形似流水从折扇流过，将折面和流线优雅地结合在一起，竖向线条使建筑形象更为挺拔。顶部通过女儿墙造型加强转折效果。裙房腰线形似淮河流水从折扇中如瀑布般流下，使建筑的形式美和意境美巧妙结合，同时也使室内空间更加生动。

Architectural design draws "Huai River Water" of the Bengbu natural landscape and "Folding Fan" intention of local culture. The plan design employs a slight twist, which not only ensures a balanced area, but also imitates the image of folding fan. Facade is shaped like water flowing on a folding fan, gracefully combining folding surface and streamline together, and its vertical lines strengthen the towering image of the hotel. At the top, turning effect is enhanced through parapet modeling. Podium waist line is designed similar to the Huai River flowing down like a waterfall from the folding fan, combining poetic imagery beauty and formal beauty and at the same time, making interior space more vivid.

INTERIOR OF HOTEL
酒店内装

LANDSCAPE OF HOTEL
酒店景观

以"蚌孕沙成珠"为设计理念，从珍珠的形成和蚌壳纹理提取设计语言，运用于景观的铺装以及小品设施当中，形成富有节奏的空间组合序列，营造丰富灵动的商业空间环境。采用"珠落玉盘"的设计手法，把景观节点化作"珍珠"，使"珍珠"形成多元化的景观空间——如主景雕塑、树池小品等——营造多种趣味的景观形式。

Following the "Pearl Forming in Clam" design philosophy, the landscape design makes its language extracted from pearl formation and clam shell texture to be visible through the pavement of landscape, small article facilities, forming a space combination sequence with rich rhythm and a vivid and rich commercial space environment. In addition, by adopting the design skill of "Pearl Falling on the Jade Plate", landscape nodes are designed as "Peal" element to present the pluralism of landscape space, such as the main sculptures, tree pool, etc., thus creating a variety of interesting landscape forms.

05

07

NIGHTSCAPE OF HOTEL
酒店夜景

05 酒店入口景观
06 酒店绿化
07 酒店内院景观
08 酒店夜景

08

WANDA
KIDSPLACE
万达宝贝王

WANDA
COMMERCIAL
PLANNING
2014

WANDA KIDS PLACE
万达宝贝王

万达儿童娱乐有限公司是万达集团旗下投资中国亲子家庭文创娱乐产业的平台公司。公司以"让孩子在梦想中成长"为经营理念，针对0~12岁亲子家庭打造中国最具影响力的乐园集群为发展目标，为中国的孩子与父母营造一个共同成长的美好空间，一个不断激发孩子梦想的欢乐世界。

万达宝贝王乐园是万达儿童娱乐有限公司首期推出的动漫亲子乐园，面积1500~4000平方米，在全国万达广场直营连锁开发。乐园专为0~12岁中国亲子家庭设计，坚持"让孩子在梦想中成长"的经营理念，通过动漫主题氛围的营造，在精彩纷呈的游乐中融入丰富多彩的体验方式，打造一个拓展儿童思维、培养儿童社交、丰富儿童生活、增进亲子情感的欢乐世界。

万达宝贝王乐园几年内将覆盖全国4百家，成为中国最大最具影响力的动漫亲子乐园。

Wanda Children Entertainment Co., Ltd is the subordinate platform company of Wanda Group targeted towards investment on cultural and creative entertainment industry for Chinese families. Insisting on the operational principle of "Dreams for Kids" and walking toward the goal of creating the China's most influential entertainment park against the children between the ages of 0~12, it seeks to offer a desirable space for parents and children to grow together, and a happy world to constantly inspire children's dreams.

Wanda Kids Place park is the carton family park firstly lunched by Wanda Children Entertainment Co., Ltd. The parks will span areas between 1,500 and 4,000 square meters and developed in the form of regular chain at Wanda Plaza nationwide. Designed especially for Chinese children between the ages of 0~12, the park insists on the operational principle of "Dreams for Kids". Through the creation of carton theme atmosphere and integration of colorful experiences into wonderful amusement, it strives to build a happy world which develops children's thoughts, cultivates children's social contact, enriches children's life and promotes family relationship.

The number of Wanda Kids Place parks will reach 400 in China in a few years, making it become China's most influential carton family park.

01

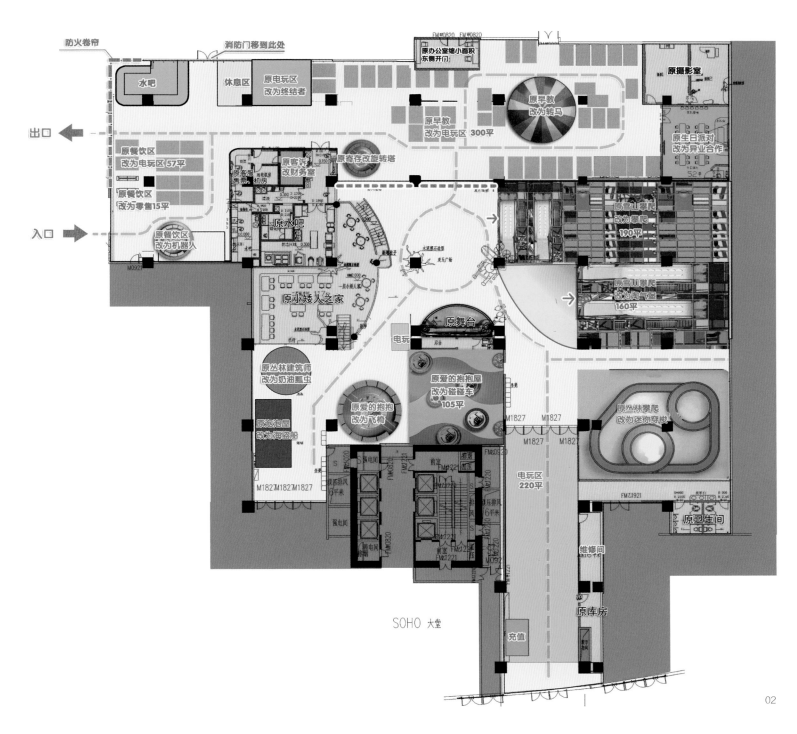

防火卷帘   消防门移到此处   FM0820 FM0820

                                            原办公室缩小面积   原摄影室
                                            乐侧开门

水吧   休息区   原电玩区
                    改为终结者   原早教
                                          改为转马

出口 ←   原生日派对
                                                            改为异业合作
原餐饮区   原客服   原客诉
改为电玩区 57平   售票   改为财务室   原寄存改旋转塔
         机房   原高山攀爬
原餐饮区   改为攀岩
改为零售15平   原水吧   190平
入口 →
         原餐饮区   原小矮人之家   原高山攀爬
         改为机器人   改为海一堂
                         未来漫石油型   欢乐广场   160平

                              原舞台
                       电玩   原丛林攀爬
         原丛林建筑师   原爱的抱抱屋   改为迷你穿梭
         改为奶油瓢虫   改为碰碰车
                              105平
         原爱的抱抱
         改为飞椅
原泡泡屋   M1827 M1827
改为海盗船   电玩区
                            220平   M1827 M1827
         M1827 M1827 M1827

                                                        原卫生间

                                           维修间

                                     电玩区
                SOHO 大堂   220平   原库房

                             充值

                                                                        02

                                                03

04

04 昆明西山万达广场宝贝王活动区
05 昆明西山万达广场宝贝王入口
06 通州万达广场宝贝王入口
07 通州万达广场宝贝平面图

06

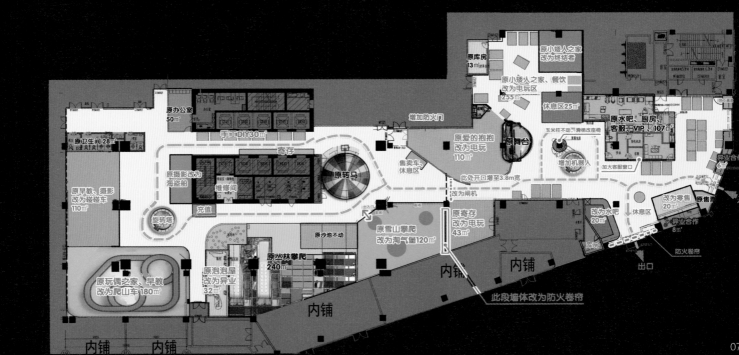

07

08

08 通州万达广场宝贝王活动场景
09 东莞东城万达广场宝贝王活动场景
10 东莞东城万达广场宝贝王活动场景
11 东莞东城万达广场宝贝王平面图

09

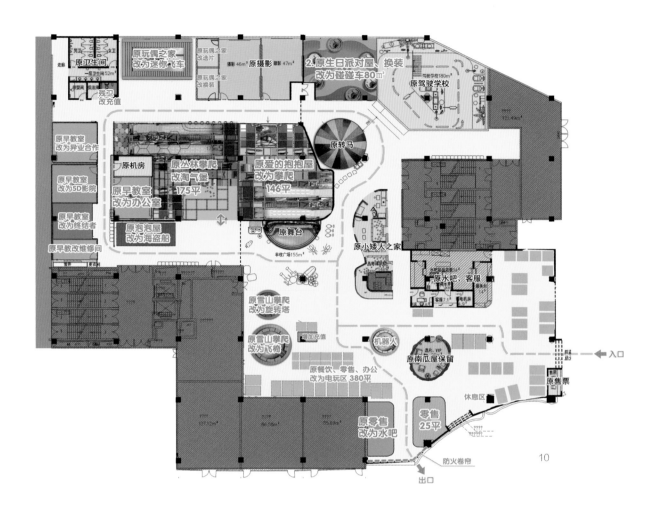

12

13

12 烟台芝罘万达广场宝贝王活动区
13 烟台芝罘万达广场宝贝王活动区
14 烟台芝罘万达广场宝贝王平面图
15 包头万达广场宝贝王平面图
16 包头万达广场宝贝王活动场景

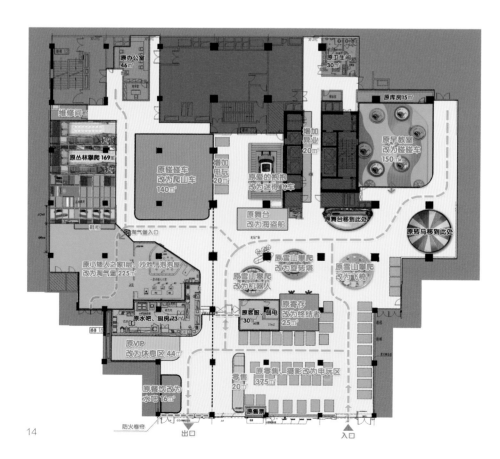

14

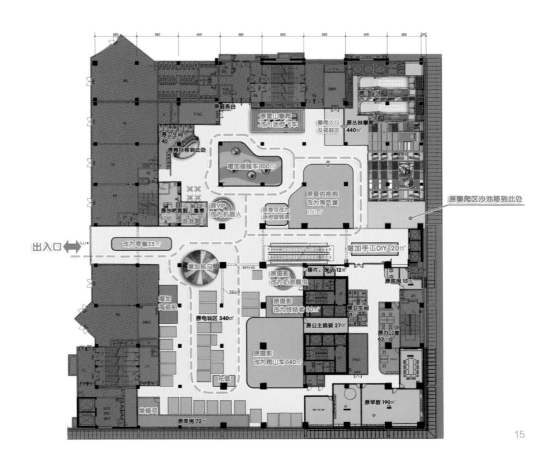

15

16

17

18

19

20

21

22

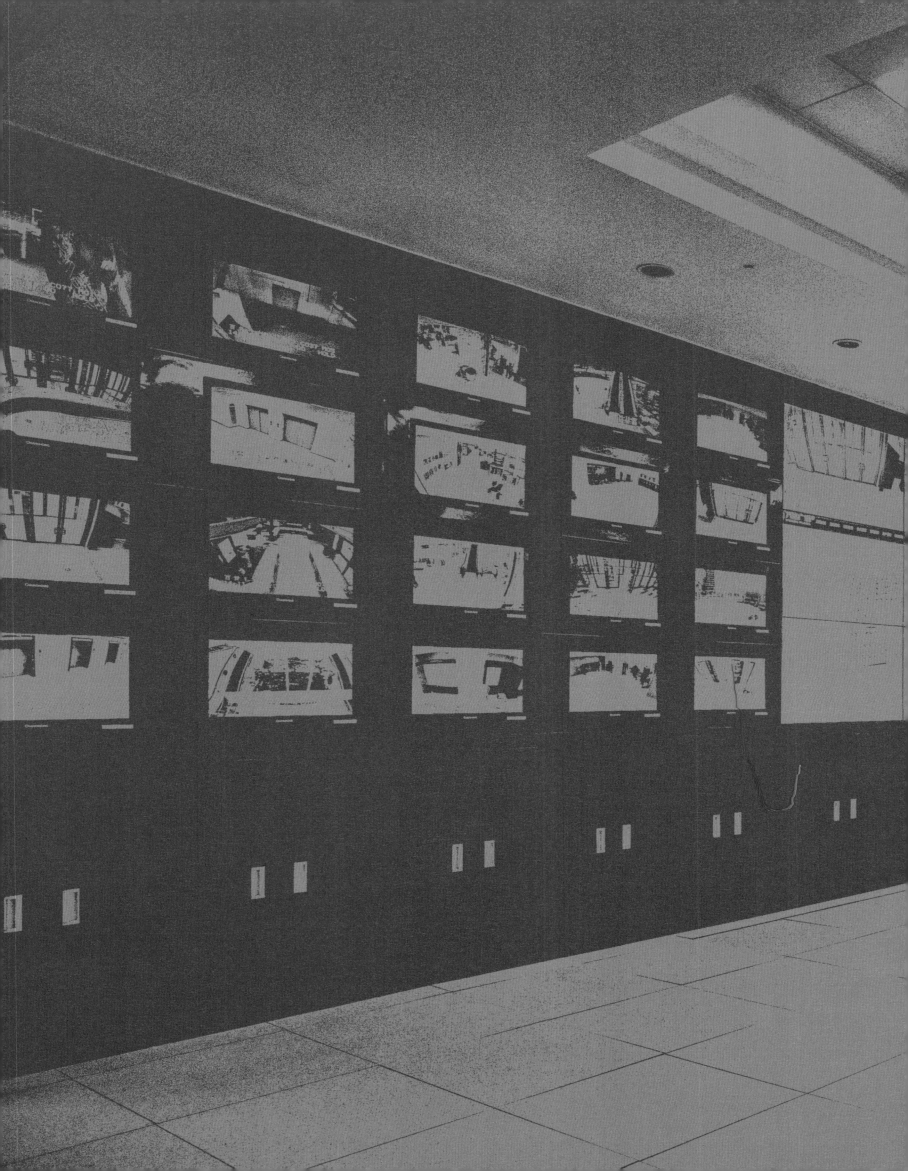

HUIYUN
INTELLIGENT
MANAGEMENT
SYSTEM
慧云智能化
管理系统

WANDA
COMMERCIAL
PLANNING
2014

ESTABLISHMENT AND DEVELOPMENT OF WANDA HUIYUN INTELLIGENT SYSTEM
万达慧云智能化系统建设与发展

万达商业规划研究院副院长　方伟

万达慧云智能化管理系统（"慧云系统"）是一种创新的商业建筑楼宇智能化监控系统。该系统将"消防管理"、"安防管理"、"设备管理"、"运营管理"和"节能管理"这"5大管理功能"和16个"弱电子系统"集成在一个操作平台上，进行集中监视、控制和管理，提高了万达广场的运营管理水平，从而实现"降低人工成本"、"保证运行品质"、"降低运行能耗"的目标。

截至2014年底，万达集团已有50个项目的慧云系统上线运行；其中包括：2013年开业的4个万达广场试点项目，2014年全部新开业24个万达广场项目和两个文化旅游项目，以及22个已开业万达广场项目的慧云系统改造。

慧云系统先后通过中国国家版权局、国家工商行政管理总局的审核，正式获得国家软件著作权和注册商标。2015年3月19日，慧云系统由中国建筑业协会智能建筑分会推荐，获得"2015年度智能建筑优质产品"奖项。万达集团一如既往地秉承了奇迹是"万达速度"，万达人也以一贯高效的执行力，引领着商业建筑智能化领域的革命。

Wanda Huiyun Intelligent Management System (Huiyun System) is a creative intelligent monitoring system for commercial buildings. With the integration of five management functions, which are fire protection management, security management, equipment management, operation management and energy saving management, and 16 weak electronic systems into one operational platform, the system makes centralized monitoring, control and management a reality, which improves Wanda's operational management competence and helps to achieve the goal of cutting down manpower costs, ensuring operating quality and lowering operating energy consumption.

By the end of 2014, Huiyun System has been implemented in 50 Wanda projects, consisting of 4 Wanda Plaza pilot projects opened in 2013, all 24 Wanda Plaza projects and 2 cultural tourism projects opened in 2014 and 22 Wanda Plaza projects that have already been under operation.

With the approval authorized successively by National Copyright Administration and National General Administration for Industry and Commerce, Huiyun System has officially acquired national software copyright and registered trademark. On 19th March, 2015, recommended by the Intelligent Building Branch of China Construction Industry Association, Huiyun System was awarded "Annual Intelligent Building High Quality Product 2015". In accordance with the marvelous "Wanda Speed", Wanda staff is leading the revolution in the field of commercial building intellectualization with Wanda's consistent highly efficient execution.

I. ESTABLISHMENT OF INSTITUTIONAL GUARANTEE FOR HUIYUN SYSTEM

In the engineering implementation stage of Huiyun Intelligent Management system, the operation is difficult and the standards are high. Involving as many as more than 10 departments of Wanda Group and more than 20 construction units, Huiyun System requires for the high attention and overall coordination provided project companies. After Huiyun System had been successfully implemented in pilot projects, the Commercial Planning Institute immediately had carried out the study and arrangement of relevant designed modules, and after discussing with the Group plan management center and various relevant departments, the Commercial Planning Institute adjusted the nodes of modules and further completed and perfected relevant standards; the adjustment and perfection towards relevant nodes, after being approved by the Group leaders, have been put into practice in the Group's system, which offers a strong guarantee for the smooth implementation of Huiyu system in future projects. (see Fig.1 and Fig.2).

II. HUIYUN SYSTEM PROJECT CONTROL AND TRAINING

In 2014, the Planning Institute took part in the whole

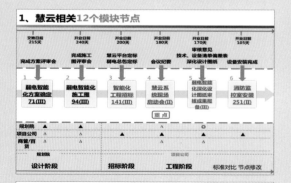

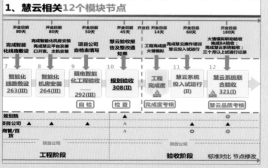

（图1）慧云模块化流程示意图

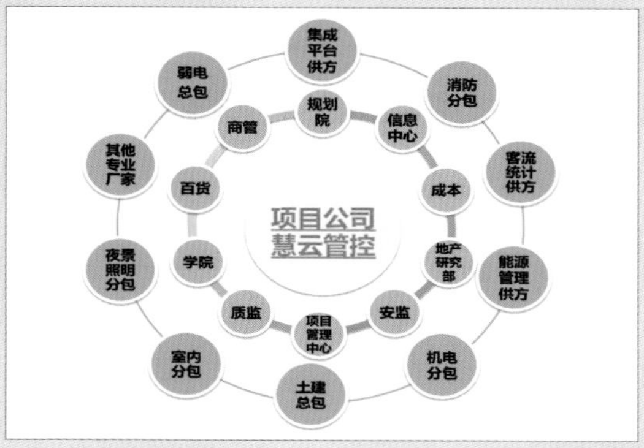

（图2）慧云涉及单位协调关系示意图

一、慧云系统建设制度保障

慧云智能化管理系统在工程实施阶段具有难度大、标准高的特点，涉及集团十余个部门、实施单位二十之多，需要项目公司高度重视、总体协调。商业规划院在试点项目慧云成功上线运行后，立即组织对相关计划模块进行梳理，与集团计划管理中心及各相关部门研讨，调整模块节点，完善完成标准；相关节点完善调整得到集团领导的签批并落实在集团制度中，有力保证了后续项目慧云系统的顺利进行（图1、图2）。

二、慧云系统项目管控与培训

2014年，规划院全程参与慧云建设的各个阶段——尤其是慧云系统启动会、慧云系统完成度考核、慧云系统联合验收等节点——付出了大量的人力，圆满完成了节点任务，起到了技术牵头部门的作用；传授慧云建设相关知识和经验，使各项目公司迅速掌握了与自身相关的慧云建设的操作方法。此外，规划院多次组织对集团各系统、各项目公司以及供方单位的相关培训工作，总结出"开好一个会、填好一个表、两次火情模拟、两次验收考核"的工作要点口诀，消除项目公司的畏难情绪，为慧云系统顺利实施创造有利条件。

三、慧云系统复盘与标准修订

2014年上半年，结合4个试点项目的使用经验，规划院联合集团多个部门，完成了慧云系统的设计、建造、验收、运行管理和培训"五个标准"的修订工作。通过此次修订，完成了慧云智能化管理系统的"四个统一"（操作界面统一、图例配色统一、硬件条件统一、机房装修统一），保证了今后建设过程中，不同的集成平台施工调试单位的成果能够保持一致，有利于商管运营管理。此外，标准增加了消防、安防系统与当地职能管理部门联网的要求，使慧云智能化管理系统成为智慧城市的一个有效的节点，再一次提高了万达购物中心的智能化水平（图3）。

process of the construction of Huiyun System, especially the launch conference of Huiyun System, the completion assessment of Huiyun System and the joint acceptance of Huiyun System. The Planning Institute made abundant efforts to successfully accomplish each task, and achieved the effects of leading other departments with technology; the Planning Institute also passed on relevant knowledge and experience concerning Huiyun System construction to project companies, which enabled project companies to rapidly grasp the operation methods for Huiyun System construction related to them. In addition, the Planning Institute organized several relevant trainings for various systems of the Group, project companies and relevant contractors of Wanda. With the summarized pithy formula of "attending one conference, filling out one chart, taking part in two fire protection simulations and two acceptance tests", the trainings successfully eliminated project companies' fear of difficulties and created favorable conditions for the successful implementation of Huiyun System.

III. CHECKING AND STANDARD REVISION OF HUIYUN SYSTEM

In the first half of 2014, in combination with the operation experience of the 4 pilot projects and with the cooperation offered by various departments of the Group, the Planning Institute completed the compilation of "Five Standards" concerning the design, construction, acceptance, operational management and training of Huiyun System. After such compilation, "Four Unifications" (unification of operation interface, unification of picture color, unification of hardware conditions and unification of machine room decoration) of Huiyun Intelligent Management System have been achieved to ensure that in the process of future construction, the achievements of integrated platform construction carried out by different construction units can be of the same standard, which shall make for the operational management of commercial management department. In addition, according to the newly added requirement in the standards, the fire protection and security system shall be connected to local functional management authorities, which shall make Huiyun Intelligent Management System an effective component of smart city and further improve the intelligence of Wanda Shopping Mall (see Fig.3).

In 2015, under the organization of the Group, special examination and check summary have been successfully completed by various business departments towards 15 Huiyun System projects under joint acceptance. Huiyun System Special Task Group was founded to re-optimize the functions and operation mode of Huiyun System, based on the problems detected in onsite inspection and checking. Besides, the standards for the design, acceptance and operation of Huiyun System were revised by relevant departments, based on which Huiyun Intelligent Management System Version 2.0 was launched.

As for design standards, in order to implement Huiyun Intelligent Management System more scientifically and reasonably and to improve the completion quality of Huiyun System, on the premise that the overall functions and one-key control of Huiyun System are not compromised, the new standards carries out optimization on the implementation models of 16 sub-systems by adopting three integration methods so as to dramatically simplify the engineering construction of Huiyun platform. In accordance with the

（图3）慧云系统1.0版标准界面

2015年，集团组织各业务部门对已完成慧云系统联合验收的15个项目进行专项检查和复盘总结，并成立慧云专项工作小组，针对现场检查和复盘发现的问题，重新优化慧云系统功能与操作模式，组织相关部门修订了慧云系统"设计、验收、运营"标准，推出慧云智能化管理系统2.0版。

在设计标准方面，为了更加科学、合理地实施慧云智能化管理系统，提高慧云系统的完成质量，新版标准在不影响慧云总体功能及一键操控的前提下，采取3种集成方式对16个子系统的功能实现方式进行优化，从而大幅降低慧云平台的工程建设难度。新版慧云系统根据使用方的实际需求，增加系统参数设定初始化、报警自定义、自定义配置视频轮巡方案、按预置模板自动生成报表、天气预报手自动获取等多项实用功能，为用户提供了更友好的人机交互体验。慧云操作台采用更符合人体工学的设计方案，并通过KVM技术，使管理人员在操作台上能够对所有弱电子系统主机进行直接操作，提升了系统整体的可靠性。

在验收标准方面，新标准引入了一票否决项和子系统及格线验收条款，并在系统投入试运行的验收节点要求进行所有点位、所有功能的全覆盖检测，从而实现对慧云系统完成质量的严格把关（图4~图6）。

四、慧云系统工作展望

新版慧云系统于2015年10月1日在新开业万达广场项目中推广实施。慧云工作小组目前正组织优秀供方开展慧云样板项目建设，并在此基础上研发慧云的"云平台"方案。慧云"云平台"方案具有云端部署、总部集中运维、统一数据备份、版本统一发布等多项优势，使慧云系统的管控及升级更加方便。此外，"云平台"方案要求各地项目数据实时上传总部，可

actual demands of the user of the new version of Huiyun System, multiple functions, including initialization of system parameters, custom alarm, custom video monitoring scheme, automatic generation of reports based on preset templates, automatic or manual acquisition of weather report, are added to provide more friendly human-computer integration experience for users. By utilizing a more ergonomic design plan and KVM technology, the administrators can directly operate the host computers of all weak current systems on the operation platform, which shall promote the overall reliability of the system.

In terms of acceptance standards, the new Standards brings in one-vote veto option and cut-off scores and clauses for sub-system acceptance; in addition, in the inspection and acceptance of a system under pilot run, detections shall cover all points and positions and all functions, so as to strictly safeguard the completion quality of Huiyun System (see Fig.4 to Fig.6).

IV. WORK PROSPECTS FOR HUIYUN SYSTEM

On 1st October, 2015, the new version of Huiyun System will be implemented in Wanda Plaza projects, which are to be opened. Currently, under the organization of Huiyun Task Group, outstanding design providers are carrying on the construction of Huiyun Sample Project, based on which the task group will develop a "Cloud Platform" Scheme. With the advantages of cloud deployment, centralized operation and maintenance by the headquarters, unified data backup and unified release of edition, the Huiyun "Cloud Platform" Scheme shall facilitate the control and upgrading of Huiyun System. In addition, the "Cloud Platform" Scheme stipulates that project companies in various regions shall upload real-time project data to the headquarters; therefore, big data analysis can be carried out based on business data, which shall create favorable conditions for the excavation of the data value in depth.

With the consistent perfection of Huiyu System, Wanda Plaza shall have more technical advantages over its competitors in the same industry in the aspect of operational management, which will not only help cut down manpower costs, ensure operating quality and reduce operating energy consumption, but also make for Wanda Plaza assets' maintenance and increase of value. With full exploitation on

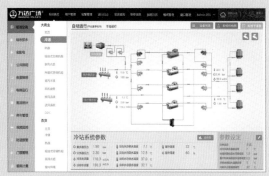

（图4）慧云2.0版-更友好的冷源操作界面

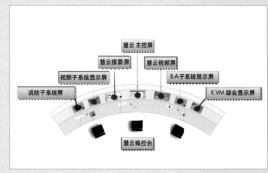

（图5）慧云操作台

（图6）慧云机房效果图

the value of the data collected by Huiyun System, favorable results of design plan further optimization, construction costs reduction, promotion of operational management ability and formation of a virtuous circle among design, construction and operation can be achieved, which shall contribute to the brand promotion of Wanda Plaza and safeguard the smooth implementation of the Group's Asset-light Strategy.

对业务数据进行大数据分析，为深度挖掘数据价值创造了有利条件。

慧云系统的不断完善，使万达广场在运营管理方面比同行业竞争者拥有更多的技术优势，有助于降低人工成本、保证运行品质、降低运行能耗，有利于万达广场资产的保值、升值。充分挖掘慧云系统的数据价值，可以进一步优化设计方案，降低建造成本，提升运营管理能力，设计、建造、运营形成良性循环，有助于万达广场品牌的推广，为集团"轻资产"战略的顺利实施保驾护航。

HUIYUN INTELLIGENT MANAGEMENT SYSTEM
慧云智能化管理系统介绍

"万达慧云智能化管理系统"，是万达集团经过自主创新研发，拥有自主知识产权的大型商业建筑智能化管理系统。该系统提供了一套全新的建筑运营一体化解决方案，通过这套系统可以实现随时随地查看并处理万达广场各项信息，比如能耗统计数据、商场客流数据、设备运行状况等。慧云智能化管理系统填补了商业综合体领域弱电智能化系统集成的空白，它的建设实施，在技术和管理上对万达集团可持续发展和高效的运营具有深远的战略意义。

"Wanda Huiyun Intelligent Management System", independently researched and developed by Wanda Group and with independent intellectual property rights, is directed at large commercial buildings. The system provides a new set of construction-operation integration solutions, through which information of Wanda Plaza can be viewed and handled anywhere at any time, such as energy consumption statistics, shopping traffic data, equipment running status, etc. The system fills the blank of ELV intelligent system integration in the field of commercial complex. It is construction and implementation have a far-reaching technically and managerially strategic significance for sustainable development and efficient operation of Wanda Group.

01

02

03

04

05

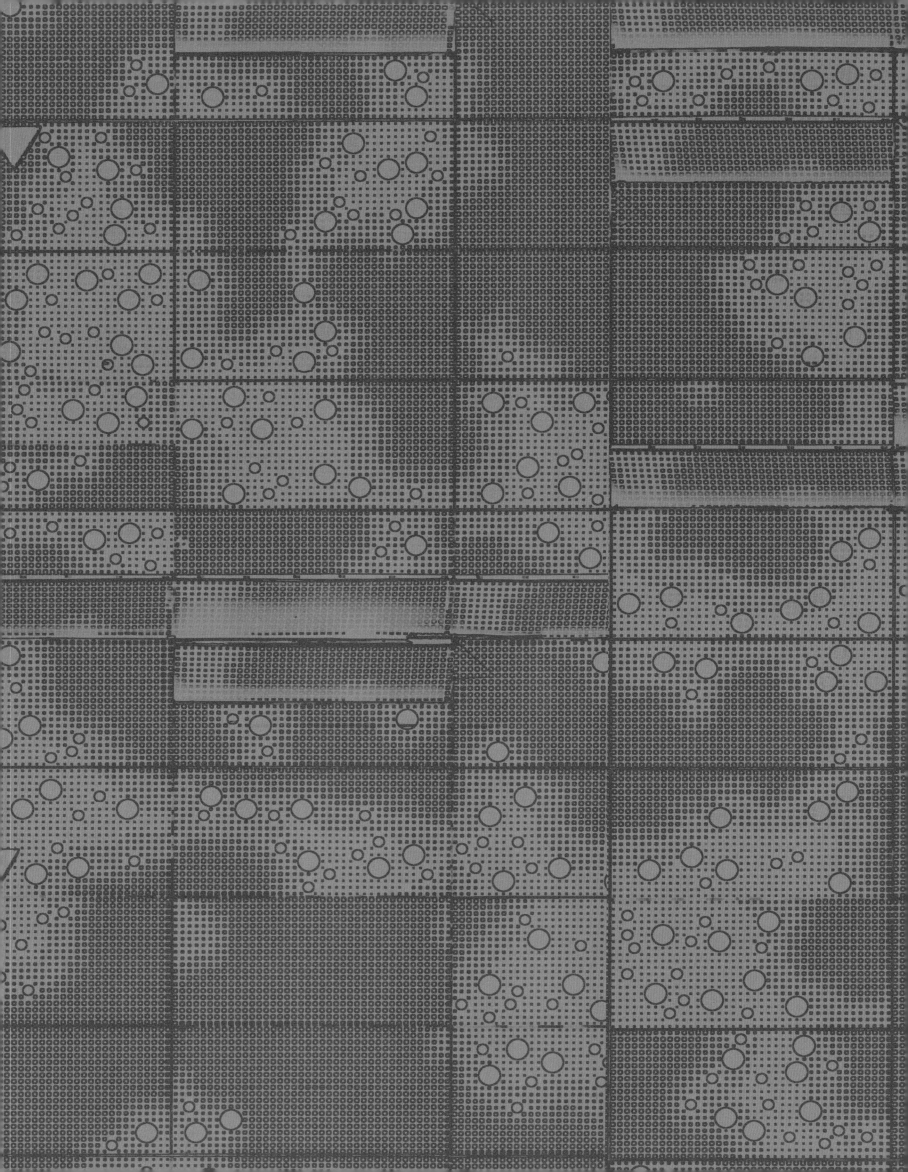

DESIGN AND CONTROL
设计与管控

WANDA
COMMERCIAL
PLANNING
2014

GLOBAL MAP OF WANDA HOTELS
万达酒店的环球版图

万达商业规划研究院总建筑师　沈文忠

一、万达酒店的源起

随着当年万达由房地产向商业地产的转型，万达开始了其酒店发展的道路。第一家万达酒店（成都索菲特）仅是现有建筑的改造项目，其后则开始了万达酒店在城市综合体中的战略布局。万达酒店可以为各万达广场建立地标效果，提升项目品质，并提供五星级餐饮、会议、婚宴、健身、客房等高质量的服务。

2012年7月万达建立了自有酒店品牌：万达瑞华、万达文华、万达嘉华；长白山国际旅游度假区项目于2012年开创了度假酒店群的先河。至2014年底，万达已拥有酒店71家，其中34家为万达自有品牌酒店（图1）。

（图1）武汉万达瑞华酒店与红灯笼秀场

二、万达酒店的发展

万达的酒店的发展经历了几个阶段：城市酒店的快速发展、文旅度假酒店群的兴起、境外酒店项目的快速扩张。

万达城市酒店快速发展，近几年每年均有17~20家酒店开业，并出现了一批高品质、多元化、各具地方文化特色的优秀城市酒店：如武汉红灯笼秀场边的顶级万达瑞华酒店，与万达百店同时开业、具有山茶花形象的昆明万达文华酒店（图2），万达第一座拥有观海空中大堂的烟台万达文华酒店，立面犹如波涛涟漪的芜湖万达嘉华酒店等。

随着万达文化旅游项目的快速发展，文旅度假酒店群也同时兴起。结合当地的人文历史、气候环境以及业态需求，各文旅项目酒店群各具特色；即使在同一

I. ORIGIN OF WANDA HOTEL

Along with Wanda's transition from real estate to commercial properties developer in those years, Wanda initiated its hotel development project. The first Wanda Hotel (Wanda Sofitel Chengdu) was merely a renovation project based on an existing building. Thereafter, Wanda Hotel started its strategic layout in City Complex. Wanda Hotel can serve as a landmark for Wanda Plaza, upgrade project quality and provide high quality services like five star accommodations, catering, conference facilities, wedding banquets, fitness center, etc.

In July 2012, Wanda has started to establish its proprietary hotel brands, including Wanda Reign, Wanda Vista and Wanda Realm; the Changbai Mountain International Tourism Resort Project in 2012 is a pioneer of resort hotel group. By the end of 2014, Wanda has already owned 71 hotels, 34 of which are hotels of Wanda's proprietary brands (see Fig.1).

II. DEVELOPMENT OF WANDA HOTEL

The development of Wanda Hotel has undergone the following stages: the rapid development of city hotel group, the rise of cultural tourism resort hotels and the rapid extension of overseas hotel projects.

With the rapid development of Wanda city hotels, 17 to 20 hotels are opened every year in recent years, some of which are outstanding city hotels with high quality, drastic varieties and local culture features, such as the high-end Wanda Reign Wuhan located near the "Red Lantern" show theatre, the camellia-imaged Wanda Vista Kunming (see Fig.2) opened at the same time with the 100th Wanda Plaza, the Wanda Vista Yantai that has Wanda's first ocean view "sky lobby", and the Wanda Realm Wuhu that has a façade in the shape of a dynamic waves pattern.

Along with the rapid development of Wanda's cultural tourism projects, the development of cultural tourism resort hotel group is also on the rise. In combination with local cultural history, climatic environment and program market demanding, each hotel group of cultural tourism projects has its own characteristic; even within the same hotel group, on the premise that the overall harmony, comprehensive momentum and complementary advantage effects are not compromised, the differentiation and uniqueness of each individual hotel are enlarged deliberately to form a resort hotel group with reasonable general layout, smooth traffic, functionally clear-cut divisions, distinct and effective streamline programs and high quality individual experience.

Classic sample of resort hotel group includes Xishuangbanna Mountain-Terrace Villa Hotel, resort hotel group of Qingdao Oriental Movie Metropolis that integrates yacht club and marina, and Wuxi resort hotel group of south China garden style.

（图2）昆明万达文华酒店

（图3）伦敦万达One项目

The acquisition of AMC cinema chain and Sunseeker Yachts in 2013 has initiated Wanda's rapid overseas development. Wanda's overseas hotel projects have also been quickly expanded in various regions all over the world. Currently, projects have been established in global metropolises like London, Madrid, Gold Coast, Chicago, Los Angeles and Sydney (see Fig. 3).

Overseas projects are distinctively diversified with strong local cultural and historical features. Because of the historical and cultural diversities of each country, the differences in local statutes and the different characteristics of local markets, each of Wanda's overseas projects has its unique features and is largely differentiated from each other. The diversified features that enable Wanda's overseas hotels to become the "destinations" network of global tourism culture, based on comprehensive and all-around experience-oriented services, and of various experience programs can be further developed.

Wanda's overseas hotels are located in the remarkable locations of world-class metropolises. Wherein, the London Hotel standing by the Thames River is like a new British aristocrat; the Madrid Spain Mansion exhibits the sedimentation and rebirth of history; the Gold Coast Jewelry Triple Towers sets off sparkles with its beauty implemented by coastal beaches; the Chicago third tallest tower will uplift the skyline of magnificent Lakefront; One Beverly Hill Wanda Vista will demonstrate the elegant demeanors of Hollywood stars; while, the One Sydney with superior ocean view will hold a panoramic scene of the Sydney Opera House and the harbor front. Taking advantages of their splendid geographic locations and mature neighborhood, and based on the analysis of city scenes from all directions, these hotels strive to maximize the utilization of scenery resources and to establish Wanda's global brand effect (see Fig. 4 to Fig. 6).

III. DESIGN CONTROL OF WANDA HOTEL

As for the Design Control of Wanda's overseas hotel projects,

酒店群中，也在追求和谐整体、综合气势和优势互补效应的基础上，刻意加大各个单体酒店的差异化和个性化，形成总图布局合理、交通顺畅、功能分区明确、流线清晰高效、具有高品质个性体验的度假酒店群。

度假酒店群的经典项目，包括西双版纳的山地别墅式酒店、青岛东方影都的海滨及游艇港结合的度假酒店群及具有江南园林风格的无锡度假酒店群等。

2013年AMC院线、圣汐游艇公司的并购，开启了万达海外的快速发展，万达境外酒店项目也在全球各地迅速展开。目前万达已经在伦敦、马德里、黄金海岸、芝加哥、洛杉矶和悉尼等全球大城市落地生根（图3）。

境外项目具有鲜明的世界多元化和地方历史文化特色。世界各国的历史文化差异、地方法规的不同、各地市场的特点，为万达各境外项目之间带来显著的差异化和个性化。正是这种差异化特点，使万达境外酒店可以发展成为环球旅游文化的"目的地"网络，进而提供多业态、全方位的综合体验式服务。

万达各境外酒店均占据着其所在大都市显著的地标位置。"伦敦酒店"有如泰晤士河畔的英伦新贵；马德里"西班牙大厦"展示着历史的沉积与再生；黄金海岸"珠宝三塔"在海岸沙滩映射下熠熠生辉；芝加哥第三高楼将更新芝加哥湖畔的天际线；"比弗利山一号"文华酒店将尽显好莱坞的明星风采；而占有无敌海景的"悉尼一号"将歌剧院和大铁桥尽收眼底。利用所在的地理位置和周边环境，通过对各方向的

（图4）黄金海岸珠宝三塔

城市景观进行分析，各境外酒店力求最大化地利用城市景观资源，打造万达的环球品牌效应（图4~图6）。

三、万达酒店的设计管控

万达酒店的设计管控，是利用国际成熟的设计市场和发达的科技水平，通过设计总包和施工总包等管控模式，借鉴万达成熟的管控制度和信息化平台进行设计管理和创新。

境外项目充分利用环球的专家资源，各项目均利用当地前五大建筑师事务所或国际明星建筑师事务所主导设计，其中包括多位获得普利兹奖的建筑大师，如诺曼·福斯特爵士、理查德·迈耶事务所等，还包括KPF、Gensler等世界上名列前茅的大型建筑师事务所。

万达酒店设计管控的另一个关键要点是资源整合。通过设计总包的模式，给予国际设计公司更大的自主性和责任感，也使其更高效地协调各专业设计分包顾问之间的配合与整合，并鼓励其创新；把当地成熟的设计方式和规范管控由境外总包设计团队完成，万达各职能部门则可以集中精力重点做好资源整合、合约管理和特色创新，进而更加有效率地管理。

万达有着一套成熟的管理制度和高度信息化管控平台。各个部门共同编写的《境外项目管理操作手册》以及《项目计划模块系统》、《项目面积填报系统》等的上线运行，使得万达的管理制度和管理模式可以运用到境外项目中，并且通过全球化的信息平台进行操作和管理（图7）。

酒店项目的设计管控也正在逐步进入全新的信息化整合平台。通过BIM（Building Information Modeling）设计手段，建立多专业、全业态整合模型，将设计、成本、计划、施工等诸多项目管控要素均整合到统一的数据化BIM模型平台中，进而达到高智能化、高科技化、高效率化的设计和项目管控（图8）。

the mature international design markets and well-developed technologies are utilized. Under the control model of Design Overall Contract and Construction Overall Contract, design management and innovation are carried out based on Wanda's mature Control System and Information Platform.

Overseas projects take full advantage of global expert resources. The leading design architect of each project are all chosen from local top-five Architects or international-claimed star Architects, some of which are Pulitzer prizes winners like Sir Norman Forster, Richard Meier & Partners Architects, and other world-class large-scale Architect firms including KPF and Gensler.

Another key point of design control of Wanda Hotel is the integration of resources. Under Design Overall Contract Model, more decision-making and responsibilities are given to overseas design consultants, which enables them to coordinate and integrate the work of sub-contractors of different specialties more efficiently and leaves them freedom for innovation. Since the design and control work are handled by experienced overseas general contract design teams, Wanda's functional departments can focus on tasks of resource integration, contract management and program innovation, and help improve the management efficiency of various departments.

Wanda has a set of mature Management System and a highly computerized Control Platform. With the execution of *Overseas Project Management Manual* compiled under the cooperation of different departments, and the implementation of *Project Schedule Module System* and *Project Area Report and Submit System*, Wanda's Management System and Management Model can be applied to overseas projects, while operations and management can be carried out on the global information platform (see Fig.7).

The design control of hotel projects is also being integrated into the new information integration platform. With the utilization of the BIM design tool, integrated models of multi-professions and complete programs are established. By integrating various contents of project management and control of design, costs, planning and construction into a unified BIM system, design and project control with high intelligence, high technology and high efficiency are achieved (see Fig.8).

IV. PROSPECT OF WANDA HOTEL

Wanda has become the world's largest luxury hotel owner with continuous development both in quantity and in quality. The global development of Wanda Hotel not only establishes Wanda's international status, but also helps to enhance Chinese culture's global influences. In addition to

（图5）马德里西班牙大厦

（图6）比弗利山一号鸟瞰图

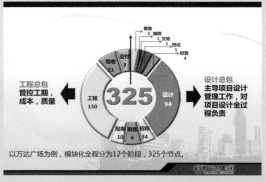

以万达广场为例，模块化全程分为12个阶段，325个节点。

（图7）境外项目325模块节点分布图

四、万达酒店的展望

万达已成为世界上最大的奢华酒店业主，不仅在数
量上，而且在品质上都有着不断的发展。万达酒店的
环球化发展不仅奠定了万达的国际化地位，而且也
可以潜移默化地推进中国文化在世界的影响力。在
强调当地人文文化和奢华服务设施外，万达酒店还
通过中餐、SPA等特性化的服务，展示现代东方空间
风格的神韵，以及"重礼仪"、重细微的个性化服务。
万达的国际化市场必将开拓出一片广阔的空间前景
和深髓的人文发展潜力，使万达成为世界超一流的
全方位服务型企业，"国际万达"必将拥有更璀璨的
明天。

（图8）芝加哥万达文华酒店

emphasize the local culture in hotel design and luxurious
service facilities, Wanda hotel can reveal the romantic charm
of modern oriental spatial-style and the personalized service
of "respective valuing" etiquette and carefulness through
featured services, for instance the Chinese restaurant and in
SPA programs. With a great cultural development potential,
Wanda will definitely open up a vast global market, which
shall pave the road for Wanda's becoming a world-class
superb service enterprise. The future for Wanda, whose goal
is to become an international enterprise with hundred years
of history, the future is certainly promising and bright.

IMPORTANCE OF LANDSCAPE CONSTRUCTION
景观在建筑中的作用

万达商业规划研究院环艺所　李斌

随着社会的发展,各式各样的建筑拔地而起,高科技的发展和现代建筑装饰材料的广泛使用,使城市空间变成了一个由钢筋混凝土围合而成的冷漠世界,使人远离自然、使城市生态环境日益恶化,极大地危害着人们的身心健康。于是我们开始审视以牺牲环境为代价的现代文明,这些真是我们引以为豪的吗?人类与自然环境的隔离和疏远是我们想要的生活吗?当然不是! 生活在城市中的居民希望亲近自然、热爱绿色,渴望把田园气息带进我们的日常周边。早在20世纪70年代"能源危机"爆发之时,就提出以"节约能源、减少污染"的理念;目前,倡导绿化、可持续发展,已经成为当今建筑设计的主题。景观设计重点是协调人与自然的关系,因此景观设计在建筑设计中的作用日趋重要。因此,景观的设计艺术如何与建筑相融合并发挥最大功效,已成为现代景观设计师们探索的重点领域。在商业建筑为主导的项目开发中,我们团队总结起来可以用三个字来概括,那就是"传"、"帮"、"带"。

All types of constructions are built along with society development. Due to the development of high technology and the extensive use of modern building and ornament materials, city space has become an indifferent world enclosed by reinforced concrete, keeping people away from the nature and causing urban ecological environment degradation, which greatly jeopardizes people's mental and physical health. Reviewing the modern civilization built at the expense of environment degradation, we have to ask ourselves whether such civilization shall be something that we are proud of. Do we really want to isolate and distance our lives from natural environment? Of course not! People living in the city yearn for nature and green and hope that a taste of rural life can be brought into their surroundings. As early as in the 1970s of the 20th century, with the outburst of "energy crisis", the concept of energy saving and pollution abatement was put forward. Currently, the advocacy for green and sustainable development has become the theme of modern architectural design. The key point of landscape design is to keep a balance between human and nature. With landscape design playing an increasingly important role in architectural design, how to integrate the art of landscape design into buildings and maximize its effect has become a key field in which modern landscape designers are exploring. In the development of projects that are mainly consisted of commercial buildings, out team has concluded three words as the tips for landscape construction, which are "reflection", "assistance" and "linkage".

一、"传"

1. 首先要传递出与建筑和谐的气质
景观要像建筑的"手和脚"一样,要传递出建筑与环境独特的气质。只有气质协调统一了,有了共同的语言,才会形成固有的环境和气场,才能整体融洽。

I. REFLECTION

1. FIRST, IT SHALL REFLECT A DISPOSITION THAT ACCORDS WITH THE BUILDINGS
Landscape shall be the "hands and foot" of buildings and shall reflect a unique disposition that accords with the

（图1）无锡万达茂景观方案

（图2）厦门集美万达广场景观方案

无锡"万达茂"建筑单体宛如一朵盛开的玉兰花，景观的总图设计和规划与玉兰花形成一幅"玉兰花千树"的美丽构图（图1）。厦门集美万达广场景观设计灵感，来源于厦门地区传统民居建筑构件和近代花桥文化影响下的厦门民居建筑元素，并结合当地著名的龙舟元素，最终达到与建筑的完美融合（图2）。

2. 其次要传承出文化

建筑的气质通过景观更进一步提高后，还需要通过景观的"手和脚"，描绘出自己的"指纹"，要有属于自己的"基因"。在每个项目之初，景观要同建筑一道，对当地及周边的博物馆——进行考察、学习和研究，将文化符号尽可能协调统一地展现在项目之中。要说建筑是气势恢宏的大手笔，那景观一定是细腻柔美的精品；而建筑与景观相容的作品才真正经得起时间沉淀的佳作。

为了做好南昌"万达茂"项目，景观与建筑负责人共同寻访景德镇的民间大师、了解景德镇的陶瓷制作过程；将制作场景加以提炼，应用于景观设计中；与建筑内部空间有效地相呼应（图3），体现了当地文化元素和趣味性（图4）。上海金山万达广场的"跳房子"、"套圈子"等情景雕塑，让人们重新回味弄堂游戏，重现童真童趣，感受老上海的情节（图5）。

二、"帮"

1. 帮忙补错

设计总是遗憾的艺术，对于建筑的一些不能弥补的遗憾，景观可以帮忙补错。景观对场地控制性较强，以景观引导的大景观规划，可以很好地协调各功能地块的相互冲突，做到形态和功能的合理结合。

buildings and surroundings. Only when the disposition is coordinated and unified can the landscape and the buildings communicate with each other, an inherent environment and aura be formed and the overall harmony be achieved.

The individual building of Wanda Mall Wuxi looks just like a blooming magnolia, and in the general drawings of landscape, the landscape together with the magnolia-shaped building form a beautiful picture of "magnolia and thousands of trees" (see Fig.1). The landscape design of Xiamen Jimei Wanda Plaza is inspired by the components of Xiamen's local traditional residential buildings and elements of Xiamen's residential buildings that are influenced by contemporary "Huaqiao Culture". In combination with the famous local dragon boat elements, the landscape is perfectly integrated into the building (see Fig.2).

2. SECOND, IT SHALL REFLECT CULTURE

After a further upgrading of its disposition by means of landscape, a building shall depict its own "fingerprint" and develop its own "DNA" through the "hands and foot" of landscape. In the initial stage of every project, personnel in charge shall visit local and surrounding museums. After inspection, study and research, they shall try their best to integrate cultural symbols into the landscape along with the building and reveal them in the projects. If we compare a building to a magnificent work, then the landscape shall be an exquisite and mellow work with high quality; and only a work that perfectly integrates building and landscape can be called an excellent work and go through the test of time.

To carry out the Nanchang Wanda Mall project well, the people in charge of landscape construction and building construction visited folk master of ceramics art in Jingdezhen together to learn the manufacturing process of Jingdezhen china; the manufacturing scenes have been extracted out and put into use in landscape design; echoing with the building's interior space (see Fig.3), it reveals the elements and interestingness of local culture (see Fig.4). In Shanghai Jinshan Wanda Plaza, there are many sculptures depicting the scenes of traditional games that people play in their childhood. These sculptures not only remind people of Nongtang games with children's innocence and delights, but also allow them to have a taste of the living scenes of Shanghai in old days (see Fig.5).

（图3）南昌万达茂景观拉坯雕塑

（图4）南昌万达茂景观情景画坯雕塑

（图5）上海金山万达广场情景雕塑

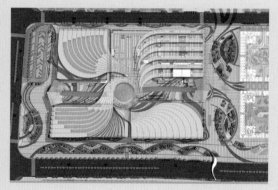

（图6）青岛万达茂景观平面图

如青岛项目，后来增加的大巴停车区域和"万达茂"主入口广场冲突，功能用地规模不足，对广场形态、车流组织及消防隔离均造成不良影响，对日后使用和管理带来隐患。通过在大景观的"统领"下，各个功能板块进行重新划分，较好地解决了这些矛盾（图6）。

2. 帮忙"补妆"

建筑有很多的附属构筑物（如风井、燃气调压站、消防栓等）裸露在环 境中，要通过景观的手法加以美化、包装，好似为建筑"补妆"。

如武汉瑞华酒店，燃气调压站按功能要求需裸露在地面，对环境造成影响。景观这时候就要帮忙"补妆"，对燃气调压站周围进行美化"包装"（图7）。

3. 帮忙补缺

虽然整体的建筑体态对场地形成较强的界定效果，但可以通过景观的调节作用，弥补建筑不能解决的遗憾。在"万达茂"的设计中景观大胆地运用了活泼的色彩、娱乐文化的元素对"万达茂"的场地品质进行了大胆的创新尝试，积极补充了建筑对场地人文关怀设施缺失的遗憾。

三、"带"

景观之于建筑一个重要的作用，就是形成了空间序列即纽带作用，起到组织并引导人们行为的作用。通过设计师的组织，以达到调动人的情绪和感受的效果。

II. ASSITANCE

1. IT SHALL ASSIST IN MAKING UP DISCREPANCY

Design is always an art of regret. In this respect, landscape can be used to make up for the regret that can't be made up by building. Landscape has a reletively strong controlling power towards sites. Therefore, by means of guiding comprehensive landscape planning with landscape, the discrepancies between different fucntional regions can be satisfactorily coordinated to achieve a reasonable connection between shape and function.

Take the Qingdao project for example, due to the inadequacy of the area of functional land, the added bus parking area forms a discrepancy with the main entrance square of "Wanda Mall", causing negative effects on square formantion, traffic flow organization and fire protection isolation, and leaving hidden danger for future use and management. Under the "leadership" of comprehensive landscape, functional modules are re-divided to solve the above discrepancies (see Fig. 6).

2. IT SHALL ASSIT IN BUILDING REFINEMENT

Many auxilliary constructions (including air shaft, gas pressure regulating station and fire hydrant) of a buiding are left exposed in the evironment. To beautify and embellish them by means of landcape is like to add make-up for the building.

Take Reign Wuhan for example, in accordance with functional requirements, the gas pressure regulating station has been left exposed on the ground, which causes bad effects on evironment. In this case, landscape is used to beautify the surroundings of the gas pressure regulating station as a make-up measure (see Fig.7).

3. IT SHALL ASSIST IN FILLING UP VACANCIES

Although the overall building modeling has a relatively strong defining effect on the site, landscape can be used to regulate and make up for the unsolved regret of a building. In the landscape design of Wanda Mall, vivid colors and elements of entertainment culture have been used, which is a bold and innovative attempt on the site quality of Wanda Mall and actively makes up for the buildings' deficiency in humanistic facilities.

III. LINKAGE

For a building, one important function of landscape is to form a spatial series, namely to serve as a "tie" that has the function of organizing and guiding people's behavior. Through the arrangement of architects, it shall achieve the effects of motivating people's emotions and feelings.

Take the Qinghai Coastal Sidewalk and Xishuangbanna A1 Road projects for example, a landscape footpath running

（图7）武汉瑞华酒店燃气调压站美化

如青岛滨海步行道及西双版纳A1路项目，通过一条贯穿项目的景观步道，将各个主要业态串联起来，真正起到了纽带的作用，使目形成统一的氛围、提升了整体品质（图8、图9）。

综上所述，景观艺术与建筑设计的融合统一是大势所趋，希望景观设计与建筑之间的"传"、"帮"、"带"愈发和谐、统一，并渴望从传统审美中寻求归属感、舒适感、亲切感。单一片面的考虑，只会羁绊社会前进的步伐；整体统一的规划设计，才可为经济社会发展和城市化进程贡献正面力量。建筑设计离不开景观的点缀融合，景观设计更需建筑的包容统一。

through the whole project connects all major structures and truly achieves its function of being a "tie", which helps establish a unified style for the project and upgrade the overall quality (see Fig.8 and Fig.9).

In conclusion, it is a trend to integrate the art of landscape into construction design. We hope that the "reflection", "assistance" and "linkage" between landscape design and buildings can become increasingly harmonious and unified, and also desire to find a sense of belonging, comfort and intimacy from traditional aesthetic. Partial and one-sided consideration will stagger the progress of society; and only unified and integrated planning can contribute positive strength to the economic and social development and the progress of urbanization. Construction design can't be separated from the embellishment and fusion of landscape, and landscape design needs to be blended into constructions.

（图8）青岛海滨步行道景观方案

（图9）西双版纳A1路方案

INDEX OF
PROJECTS
项目索引

WANDA
COMMERCIAL
PLANNING
2014

INDEX OF
WANDA PLAZAS
万达广场索引

KUNMING XISHAN
WANDA PLAZA
昆明西山万达广场

大商业施工图设计单位	中国电子工程设计院
外立面设计单位	晋思建筑咨询（上海）有限公司上海分公司
内装设计单位	中艺建筑装饰有限公司
景观设计单位	华汇工程设计集团股份有限公司
导向标识设计单位	北京广育德视觉技术股份有限公司
夜景照明设计单位	深圳市标美照明设计工程有限公司
弱电智能化设计单位	北京国安电气有限公司
外幕墙深化设计单位	北京市金星卓宏幕墙工程有限公司

DONGGUAN
DONGCHENG
WANDA PLAZA
东莞东城万达广场

大商业施工图设计单位	深圳华森建筑与工程设计顾问有限公司
外立面设计单位	思邦建筑设计咨询（上海）有限公司北京分公司
内装设计单位	北京清尚艺术建筑设计院有限公司
	中深建装饰设计工程有限公司
景观设计单位	宝佳丰（北京）国际建筑景观规划有限公司
导向标识设计单位	北京艺同博雅企业形象设计有限公司
夜景照明设计单位	北京三色石环境艺术设计有限公司
弱电智能化设计单位	上海智信世创智能系统集成有限公司
外幕墙深化设计单位	厦门开联装饰工程有限公司

GUANGZHOU
ZENGCHENG
WANDA PLAZA
广州增城万达广场

大商业施工图设计单位	广州市城市规划勘测设计研究院
	广州宝贤华瀚建筑工程设计有限公司
外立面设计单位	华凯派特建筑设计（上海）有限公司
	上海汉米敦建筑设计有限公司
内装设计单位	北京清尚环艺建筑设计院有限公司
景观设计单位	北京中建建筑设计院有限公司上海分公司
导向标识设计单位	北京艺同博雅企业形象设计有限公司
夜景照明设计单位	北京三色石环境艺术设计有限公司
外幕墙深化设计单位	北京金星卓宏幕墙工程有限公司

WEIFANG
WANDA PLAZA
潍坊万达广场

大商业施工图设计单位	青岛腾远建筑设计事务所
外立面设计单位	青岛北洋建筑设计有限公司
内装设计单位	中国建筑设计研究院
景观设计单位	宝佳丰（北京）国际建筑景观规划设计有限公司
导向标识设计单位	北京清华城市规划设计研究院
夜景照明设计单位	北京市三色石环境艺术设计有限公司
外幕墙深化设计单位	北京市金星卓宏幕墙工程有限公司

SHANGHAI
SONGJIANG
WANDA PLAZA
上海松江万达广场

大商业施工图设计单位	中国建筑上海设计研究院有限公司
外立面设计单位	上海汉米敦建筑设计有限公司
内装设计单位	北京城建长城建筑装饰工程有限公司
景观设计单位	北京中建建筑设计院有限公司上海分公司
导向标识设计单位	北京视域四维城市导向系统规划设计有限公司
夜景照明设计单位	深圳市千百辉照明工程有限公司
弱电智能化设计单位	上海智信世创智能集成有限公司
外幕墙深化设计单位	深圳蓝波幕墙及光伏工程有限公司

CHIFENG
WANDA PLAZA
赤峰万达广场

大商业施工图设计单位	北京东方国兴建筑设计有限公司
外立面设计单位	北京东方国兴建筑设计有限公司
内装设计单位	北京城建长城建筑装饰工程有限公司
景观设计单位	北京中建建筑设计院有限公司上海分公司
导向标识设计单位	北京欧德标识制造有限公司
夜景照明设计单位	深圳市千百辉照明工程有限公司
弱电智能化设计单位	大连理工科技有限公司
外幕墙深化设计单位	厦门开联装饰工程有限公司

JINING TAIBAI ROAD
WANDA PLAZA
济宁太白路万达广场

大商业施工图设计单位	北京市建筑设计研究院有限公司
外立面设计单位	华凯国际（香港有限公司）
内装设计单位	深圳市三九装饰工程有限公司
景观设计单位	宝佳丰（北京）国际建筑景观规划设计有限公司
导向标识设计单位	北京视域四维城市导向系统规划设计有限公司
夜景照明设计单位	深圳普莱思照明设计顾问有限责任公司
外幕墙深化设计单位	北京市金星卓宏幕墙工程有限公司

JINHUA
WANDA PLAZA
金华万达广场

大商业施工图设计单位	华汇工程设计集团股份有限公司
外立面设计单位	青岛北洋建筑设计有限公司
	亨派建筑方案咨询（上海）有限公司
内装设计单位	中艺（北京）建筑设计研究院有限公司
景观设计单位	泛亚景观设计（上海）有限公司
导向标识设计单位	北京艺同博雅企业形象设计有限公司
夜景照明设计单位	北京三色石环境艺术设计有限公司
弱电智能化设计单位	上海中电子系统工程有限公司
外幕墙深化设计单位	厦门开联装饰工程有限公司

CHANGZHOU WUJIN
WANDA PLAZA
常州武进万达广场

大商业施工图设计单位	江苏筑森建筑设计有限公司
外立面设计单位	北京华雍汉维建筑咨询有限公司
内装设计单位	北京市建筑装饰设计院有限公司
景观设计单位	上海帕莱登建筑景观咨询有限公司
导向标识设计单位	北京视域四维城市导向系统规划设计有限公司
夜景照明设计单位	北京三色石环境艺术设计有限公司
外幕墙深化设计单位	北京市金星卓宏幕墙工程有限公司

FOSHAN NANHAI
WANDA PLAZA
佛山南海万达广场

大商业施工图设计单位	悉地国际设计顾问（深圳）有限公司
外立面设计单位	上海霍普建筑设计事务所有限公司
内装设计单位	北京城建长城建筑装饰工程有限公司
景观设计单位	上海帕莱登建筑景观咨询有限公司
导向标识设计单位	北京视域四维城市导向系统规划设计有限公司
夜景照明设计单位	北京鱼禾光环境设计有限公司
弱电智能化设计单位	深圳市顺恒利科技工程有限公司
	北京益泰牡丹电子工程有限责任公司
外幕墙深化设计单位	深圳蓝波绿建与幕墙有限公司
	上海旭密林幕墙有限公司

MA'ANSHAN
WANDA PLAZA
马鞍山万达广场

大商业施工图设计单位	安徽省建筑设计研究院有限责任公司
外立面设计单位	上海思亚建筑设计咨询有限公司
内装设计单位	北京清尚环艺建筑设计有限公司
景观设计单位	上海兴田建筑工程设计事务所
导向标识设计单位	北京艺同博雅企业形象设计有限公司
夜景照明设计单位	深圳普莱思照明设计顾问有限责任公司
弱电智能化设计单位	上海中电电子系统工程有限公司
外幕墙深化设计单位	上海旭密林幕墙有限公司

JINGZHOU
WANDA PLAZA
荆州万达广场

大商业施工图设计单位	中信建筑设计研究总院有限公司
外立面设计单位	北京赫斯科建筑设计咨询有限公司
内装设计单位	北京清尚环艺建筑设计有限公司
景观设计单位	上海帕莱登建筑景观咨询有限公司
导向标识设计单位	北京艺同博雅企业形象设计有限公司
夜景照明设计单位	深圳金达照明工程有限公司
弱电智能化设计单位	北京益泰牡丹电子工程有限责任公司
外幕墙深化设计单位	深圳蓝波幕墙及光伏工程有限公司

BEIJING TONGZHOU WANDA PLAZA
北京通州万达广场

大商业施工图设计单位　　北京维拓时代建筑设计有限公司
外立面设计单位　　　　　思邦建筑设计咨询（上海）有限公司
内装设计单位　　　　　　北京城建长城建筑装饰工程有限公司 &ILYA
景观设计单位　　　　　　宝佳丰（北京）国际建筑景观规划设计有限公司
导向标识设计单位　　　　北京广育德视觉技术股份有限公司
夜景照明设计单位　　　　深圳市千百辉照明工程有限公司
弱电智能化设计单位　　　北京益泰牡丹电子工程有限公司
外幕墙深化设计单位　　　中国建筑科学研究院

NANNING QINGXIU WANDA PLAZA
南宁青秀万达广场

大商业施工图设计单位　　上海现代建筑设计（集团）有限公司
外立面设计单位　　　　　上海汉米敦建筑设计有限公司
内装设计单位　　　　　　北京清尚环艺建筑设计院有限公司
景观设计单位　　　　　　上海帕莱登建筑景观咨询有限公司
导向标识设计单位　　　　北京视域四维城市导向系统规划设计有限公司
夜景照明设计单位　　　　深圳市金达照明股份有限公司
弱电智能化设计单位　　　上海智信世创智能集成有限公司
外幕墙深化设计单位　　　北京金星卓宏幕墙工程有限公司

LANZHOU CHENGGUAN WANDA PLAZA
兰州城关万达广场

大商业施工图设计单位　　北京东方国兴建筑设计院有限公司
外立面设计单位　　　　　上海帕莱登建筑景观咨询有限公司
内装设计单位　　　　　　北京城建长城建筑装饰设计工程有限公司
景观设计单位　　　　　　华东建筑设计研究院有限公司
导向标识设计单位　　　　北京艺同博雅企业形象设计有限公司
夜景照明设计单位　　　　深圳市凯铭电气照明有限公司
弱电智能化设计单位　　　大连理工科技有限公司
外幕墙深化设计单位　　　开联建工（集团）有限公司

LONGYAN WANDA PLAZA
龙岩万达广场

大商业施工图设计单位　　中国建筑西南建筑设计院
外立面设计单位　　　　　北京赫斯科建筑设计咨询有限公司
　　　　　　　　　　　　亚瑞建筑设计有限公司
内装设计单位　　　　　　北京市建筑装饰设计院有限公司
景观设计单位　　　　　　北京中建建筑设计院有限公司上海分公司
导向标识设计单位　　　　北京视域四维城市导向系统规划设计有限公司
夜景照明设计单位　　　　深圳市千百辉照明工程有限公司
外幕墙深化设计单位　　　北京市金星卓宏幕墙工程有限公司

GUANGZHOU PANYU WANDA PLAZA
广州番禺万达广场

大商业施工图设计单位　　广州市城市规划勘测设计研究院
外立面设计单位　　　　　上海鼎实建筑设计有限公司
内装设计单位　　　　　　广东省集美设计工程公司
景观设计单位　　　　　　广州市东篱环境艺术有限公司
导向标识设计单位　　　　北京艺同博雅企业形象设计有限公司
夜景照明设计单位　　　　深圳市标美照明设计工程有限公司
弱电智能化设计单位　　　华体集团有限公司
外幕墙深化设计单位　　　北京金星卓宏幕墙工程有限公司

YANTAI ZHIFU WANDA PLAZA
烟台芝罘万达广场

大商业施工图设计单位　　青岛北洋建筑设计有限公司
外立面设计单位　　　　　上海汉米敦建筑设计有限公司
内装设计单位　　　　　　北京城建长城建筑装饰设计工程有限公司
景观设计单位　　　　　　上海兴田建筑工程设计事务所
导向标识设计单位　　　　北京艺同博雅企业形象设计有限公司
夜景照明设计单位　　　　北京鱼禾光环境有限公司
弱电智能化设计单位　　　华体集团有限公司
外幕墙深化设计单位　　　深圳蓝波绿建集团股份有限公司

JIANGMEN WANDA PLAZA
江门万达广场

大商业施工图设计单位　　悉地国际设计顾问（深圳）有限公司
外立面设计单位　　　　　上海帕莱登建筑景观咨询有限公司
内装设计单位　　　　　　上海帕莱登建筑景观咨询有限公司
景观设计单位　　　　　　上海帕莱登建筑景观咨询有限公司
导向标识设计单位　　　　北京视域四维城市导向系统规划设计有限公司
夜景照明设计单位　　　　北京鱼禾光环境设计有限公司
外幕墙深化设计单位　　　厦门开联装饰工程有限公司

FUQING WANDA PLAZA
福清万达广场

大商业施工图设计单位　　深圳奥意建筑工程设计有限公司
外立面设计单位　　　　　华凯建筑设计（上海）有限公司
内装设计单位　　　　　　深圳三九装饰工程有限公司
景观设计单位　　　　　　华汇工程设计集团股份有限公司
导向标识设计单位　　　　北京艺同博雅企业形象设计有限公司
夜景照明设计单位　　　　栋梁国际照明设计（北京）中心有限公司
弱电智能化设计单位　　　上海中电电子信息工程有限公司
外幕墙深化设计单位　　　深圳蓝波幕墙及光伏工程有限公司

WENZHOU PINGYANG WANDA PLAZA
温州平阳万达广场

大商业施工图设计单位　　华汇工程设计集团股份有限公司
外立面设计单位　　　　　上海欧创商务咨询服务有限公司
内装设计单位　　　　　　北京清尚环艺建筑设计院有限公司
景观设计单位　　　　　　中国建筑设计有限公司
导向标识设计单位　　　　北京艺同博雅企业形象设计有限公司
夜景照明设计单位　　　　北京三色石环境艺术设计有限公司
弱电智能化设计单位　　　上海智信世创智能系统集成有限公司
外幕墙深化设计单位　　　上海旭密林幕墙有限公司

HANGZHOU GONGSHU WANDA PLAZA
杭州拱墅万达广场

大商业施工图设计单位　　华汇工程设计集团股份有限公司
外立面设计单位　　　　　上海鼎实建筑设计有限公司
内装设计单位　　　　　　北京清尚环艺建筑设计院有限公司
景观设计单位　　　　　　上海帕莱登建筑景观咨询有限公司
导向标识设计单位　　　　北京视域四维城市导向系统规划设计有限公司
夜景照明设计单位　　　　栋梁国际照明设计（北京）中心有限公司
弱电智能化设计单位　　　大连理工科技有限公司
外幕墙深化设计单位　　　厦门开联装饰工程有限公司

YINCHUAN XIXIA WANDA PLAZA
银川西夏万达广场

大商业施工图设计单位　　中国电子工程设计院
外立面设计单位　　　　　华凯建筑设计（上海）有限公司
内装设计单位　　　　　　北京城建长城建筑装饰工程有限公司
　　　　　　　　　　　　北京丽贝亚建筑装饰工程有限公司
景观设计单位　　　　　　华汇工程设计集团股份有限公司
导向标识设计单位　　　　北京艺同博雅企业形象设计有限公司
夜景照明设计单位　　　　深圳市标美照明设计工程有限公司
外幕墙深化设计单位　　　深圳蓝波绿建集团股份有限公司

MANZHOULI WANDA PLAZA
满洲里万达广场

大商业施工图设计单位　　哈尔滨工业大学建筑设计研究院
外立面设计单位　　　　　上海帕莱登建筑景观咨询有限公司
内装设计单位　　　　　　北京城建长城建筑装饰工程有限公司
景观设计单位　　　　　　宝佳丰（北京）国际建筑景观规划设计有限公司
导向标识设计单位　　　　北京视域四维城市导向系统规划设计有限公司
夜景照明设计单位　　　　上海易照景观设计有限公司
弱电智能化设计单位　　　大连理工科技有限公司
外幕墙深化设计单位　　　深圳蓝波绿建集团股份有限公司

INDEX OF
WANDA HOTELS
万达酒店索引

**WANDA VISTA
KUNMING**
昆明万达文华酒店

施工图设计单位	广东省建筑设计研究院（深圳分院）
外立面设计单位	晋思建筑设计咨询（上海）有限公司
内装设计单位	万达酒店设计院
景观设计单位	华汇工程设计集团股份有限公司
夜景照明设计单位	深圳市标美照明设计工程有限公司
外幕墙深化设计单位	北京市金星卓宏幕墙工程有限公司

**WANDA VISTA
YANTAI**
烟台万达文华酒店

施工图设计单位	青岛北洋建筑设计有限公司
外立面设计单位	上海汉米敦建筑设计有限公司
内装设计单位	北京城建长城建筑装饰工程有限公司
景观设计单位	上海兴田建筑工程设计事务所
夜景照明设计单位	北京鱼禾光环境有限公司
外幕墙深化设计单位	上海蓝波绿建集团股份有限公司

**WANDA VISTA
DONGGUAN**
东莞万达文华酒店

施工图设计单位	深圳华森建筑与工程设计顾问有限公司
外立面设计单位	思邦建筑设计咨询（上海）有限公司
内装设计单位	LEOINTER
景观设计单位	宝佳丰（北京）国际建筑景观规划设计有限公司
夜景照明设计单位	北京三色石环境艺术设计有限公司
外幕墙深化设计单位	厦门开联装饰工程有限公司

**WANDA REALM
GUANGZHOU
ZENGCHENG**
广州增城万达嘉华酒店

施工图设计单位	广州宝贤华翰建筑工程设计有限公司
外立面设计单位	上海汉米敦建筑设计有限公司
内装设计单位	君衡装饰设计工程（上海）有限公司
景观设计单位	北京中建建筑设计院有限公司上海分公司
夜景照明设计单位	北京三色石环境艺术设计有限公司
外幕墙深化设计单位	北京金星卓宏幕墙工程有限公司

**WANDA PULLMAN
WEIFANG**
潍坊万达铂尔曼酒店

施工图设计单位	青岛腾远设计事务所有限公司
外立面设计单位	青岛北洋建筑设计有限公司
内装设计单位	J&A 姜峰室内设计有限公司
景观设计单位	上海帕莱登建筑景观咨询有限公司
夜景照明设计单位	北京三色石环境艺术设计有限公司
外幕墙深化设计单位	北京金星卓宏幕墙工程有限公司

**WANDA REALM
CHIFENG**
赤峰万达嘉华酒店

施工图设计单位	北京东方国兴建筑设计有限公司
外立面设计单位	北京东方国兴建筑设计有限公司
内装设计单位	香港郑中设计事务所（CCD）
景观设计单位	上海帕莱登建筑景观咨询有限公司
夜景照明设计单位	深圳千百辉照明工程有限公司
外幕墙深化设计单位	厦门开联装饰有限公司

**WANDA REALM
JINING**
济宁万达嘉华酒店

施工图设计单位	北京市建筑设计研究院
外立面设计单位	华凯国际（香港有限公司）
内装设计单位	万达酒店设计院
景观设计单位	宝佳丰（北京）国际建筑景观规划设计有限公司
夜景照明设计单位	深圳普莱思照明设计顾问有限责任公司
外幕墙深化设计单位	北京市金星卓宏幕墙工程有限公司

**WANDA REALM
JINHUA**
金华万达嘉华酒店

施工图设计单位	华汇工程设计集团股份有限公司
外立面设计单位	亨派建筑方案咨询有限公司（上海）HPP
内装设计单位	香港郑中设计事务所（CCD）
景观设计单位	泛亚景观设计（上海）有限公司
夜景照明设计单位	北京三色石环境艺术设计有限公司
外幕墙深化设计单位	厦门开联装饰工程有限公司

**WANDA REALM
CHANGZHOU**
常州万达嘉华酒店

施工图设计单位	江苏筑森建筑设计有限公司
外立面设计单位	豪斯泰勒张思图德建筑设计咨询（上海）有限公司
内装设计单位	万达酒店设计院
景观设计单位	上海帕莱登建筑景观咨询有限公司
夜景照明设计单位	北京三色石环境艺术设计有限公司
外幕墙深化设计单位	北京市金星卓宏幕墙工程有限公司

**WANDA REIGN
WUHAN**
武汉万达瑞华酒店

施工图设计单位	广州珠江外资建筑设计院
外立面设计单位	美艾克建筑设计咨询（北京）有限公司
内装设计单位	万达酒店设计研究院室内设计一所
景观设计单位	艺普得城市设计咨询公司
夜景照明设计单位	栋梁国际照明设计（北京）中心有限公司
外幕墙深化设计单位	华纳工程咨询（北京）有限公司

**WANDA VISTA
NANNING**
南宁万达文华酒店

施工图设计单位	上海现代建筑设计（集团）有限公司
外立面设计单位	上海汉米敦建筑设计有限公司
内装设计单位	万达酒店设计院
景观设计单位	上海帕莱登建筑景观咨询有限公司
夜景照明设计单位	深圳市金达照明股份有限公司
外幕墙深化设计单位	北京市金星卓宏幕墙工程有限公司

**WANDA VISTA
LANZHOU**
兰州万达文华酒店

施工图设计单位	东方国兴建筑设计有限公司
外立面设计单位	上海帕莱登建筑景观咨询公司
内装设计单位	深圳市黎奥室内设计有限公司
景观设计单位	华东建筑设计研究院有限公司
夜景照明设计单位	深圳市凯奥电气照明有限公司
外幕墙深化设计单位	厦门开联装饰工程有限公司

**WANDA REALM
MA'ANSHAN
马鞍山万达嘉华酒店**

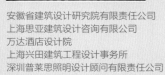

施工图设计单位	安徽省建筑设计研究院有限责任公司
外立面设计单位	上海思亚建筑设计咨询有限公司
内装设计单位	万达酒店设计院
景观设计单位	上海兴田建筑工程设计事务所
夜景照明设计单位	深圳普莱思照明设计顾问有限责任公司
外幕墙深化设计单位	上海旭密林幕墙有限公司

**WANDA REALM
JINGZHOU
荆州万达嘉华酒店**

施工图设计单位	中信建筑设计研究总院有限公司
外立面设计单位	上海思亚建筑设计咨询有限公司
内装设计单位	万达酒店设计院
景观设计单位	上海帕莱登建筑景观咨询有限公司
夜景照明设计单位	深圳市金达照明股份有限公司
外幕墙深化设计单位	深圳蓝波幕墙及光伏工程有限公司

**WANDA REALM
LONGYAN
龙岩万达嘉华酒店**

施工图设计单位	中国中轻国际工程有限公司
外立面设计单位	亚瑞建筑设计有限公司
内装设计单位	万达酒店设计院
景观设计单位	北京中建建筑设计院有限公司上海分公司
夜景照明设计单位	深圳市千百辉照明工程有限公司
外幕墙深化设计单位	北京市金星卓宏幕墙工程有限公司

**WANDA REALM
JIANGMEN
江门万达嘉华酒店**

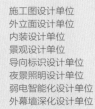

施工图设计单位	悉地国际设计顾问（深圳）有限公司
外立面设计单位	上海帕莱登建筑景观咨询有限公司
内装设计单位	万达酒店设计院
景观设计单位	上海帕莱登建筑景观咨询有限公司
夜景照明设计单位	北京万德门特城市照明设计有限公司
外幕墙深化设计单位	厦门开联装饰工程有限公司

**WANDA REALM
WUHU
芜湖万达嘉华酒店**

施工图设计单位	中国建筑上海设计研究院有限公司
外立面设计单位	北京华雍汉维建筑咨询有限公司
内装设计单位	万达酒店设计院
景观设计单位	福建泛亚远景环境设计工程有限公司
导向标识设计单位	大连依斯特图文导视设计工程有限公司
夜景照明设计单位	深圳市凯铭电气照明有限公司
弱电智能化设计单位	万达酒店设计研究院有限公司
外幕墙深化设计单位	深圳蓝波幕墙及光伏工程有限公司

**WANDA REALM
BENGBU
蚌埠万达嘉华酒店**

施工图设计单位	安徽省建筑设计研究院有限公司
外立面设计单位	青岛腾远设计事务所有限公司
内装设计单位	万达酒店设计院
景观设计单位	深圳文科园林股份有限公司
导向标识设计单位	北京视域四维城市导向系统规划设计有限公司
夜景照明设计单位	深圳市金达照明股份有限公司
弱电智能化设计单位	万达酒店设计研究院有限公司
外幕墙深化设计单位	北京金星卓宏幕墙工程有限公司

万达商业规划

持有类物业 下册 VOL.2

WANDA COMMERCIAL PLANNING 2014 PROPERTIES FOR HOLDING

2014

朱其玮 吴绿野 王群华 叶宇峰 任志忠 兰勇 兰峻文 张涛 黄勇
赖建燕 孙培宇 张琳 曹亚星 吴晓璐 苗凯峰 徐小莉 尚海燕
李文娟 刘婷 安竞 马红 曹春 侯卫华 张振宇 范珑 杨旭
谷建芳 张振宇 李淑仪 叶甲刚 雷磊 王鑫 李彬 张鹤 张飚
毛晓虎 莫鑫 都晖 刘江 蓝毅 郝宁克 屈娜 冯腾飞 张宝鹏
邵汀潇 万志斌 孙佳宁 袁志浩 阎红伟 吴迪 李斌 徐立军
宫赫谣 王雪松 张立峰 陈维 谢冕 刘杰 党恩 高振江
孙海龙 沈余 李昕 李海龙 罗沁 周澄 孙辉 齐宗新 黄引达
刘冰 杨艳坤 潘立影 程欢 邓金坷 康斌 刘易昆 李浩然
李江涛 钟光辉 张宁 黄春林 黄国辉 张洋 石路也 孟祥宾
刘阳 刘佩 耿大治 章宇峰 陈杰 冯志红 谷强 李小强 葛宁
张鹏翔 虞朋 田中 李洪涛 吕鲲 康宇 王治天 朱岩 董根泉
任腾飞 王吉 沈文忠 张珈博 张震 刘洋 胡存珊 马逸均
李光 郭晨光 朱迪 王锋 谢杰 李志华 宋锦华 刘锋
方文奇 秦鹏华 杨东 张堃 凌峰 李涛 张宇 易帆 任洪生
李明泽 陈勇 刘刚 刘宵 郭雪峰 赵洪斌 孔新国 陈嘉 王玉龙
刘志业 李浩 冯董 黄路 曹彦斌 周德 张剑锋 肖敏 李易 段堃
闫颇 朱欢 唐杰 刘潇 张雪晖 熊厚 王静 黄川东 董明海 王凡
谢云 王昉 黄涛 锡锋 李捷 解放 庞博 关发扬 赵青扬
任意刚 张争 张志斌 辛欣 罗贤君 郭杨 傅博 李梦雷 赵陨
杨春龙 路滨 张顺 王少雷 汪家绍 顾梦炜 姜云娇 江智亮
白宝伟 王凤华 贺明 李健 卫立新 冯晓芳 庞庆 何志勇
宋永成 谭訸 郭宇飞 高杉楠 卜少乐 刘海洋 韩冰 高峰
王睿麟 王凯 王宝柱 野天星 王瑶 葛朗 张佳 王晓 徐春辉
王永磊 李常春 曹国峰 于崇 张翺 杨汉国 赵剑利 王文广
张永战 李晓山 罗冲 王权 张旭 赵晓萌 高达 方伟 刘俊
康冠军 陈海亮 晁志鹏 邹洪 郑鑫 周永会 陈志强 桑伟
张德志 陈涛 张宇楠 高霞 王清文 王俊君 吴凡 张黎明 谭瑶
张绍哲 汤英杰 全永强 钱昆 路清淇 刘安 林彦 康兴梁
陈晓州 白宇 白夜 周明 崔勇 陈理力 杨娜 杨华 韦云 马辉
刘昕 金柱 王朝忠 罗琼 洪斌 赵宁 刘晓波 韩博 张烁君
徐广揆 魏大强 金博 马骁 程波 王鹏 柏久绪 闵盛勇 朱广宇
蒲峰 杜晶晶 汤钧 主佳 张浩 李扬 孟昆廷 赵海滨 钟文渊
王云 王奕 梁超 李丽 张雁翔 余斌 陈玭潭 韩天 宋雷 张述杰
王进纯 马雪健 李达 李万勇 耿磊 王政 王翔 张啸鸿
马长宁 姚建刚 万勇 李韦达 杨琳 马刚 王连发 殷超 刘向阳
李典 曹羽 陶晓晨 李民伟 张晓冬 法尔科内·马利亚 李华
卡斯特罗·索菲娅 季文君 刘怡德 夏海青 诺贝·马科斯 马佳
李云 张伟 孙穆元 吴科帆 武振衡 赵良颖 罗强 弓永康
廖伟 赵明 冯科力 张德志 陆宇亮 卫婷 林涛 曹玲妹
马嘉岳 柴刚军 孟晗 栾海 陆峰 林彬 王宇石 赵旭千

（以入职先后为序）

萬達吾業規畫